LEARNING FROM NORMAL WORK

HOW TO REDUCE RISK WHEN NOTHING GOES WRONG

DR MARCIN NAZARUK

Learning From Normal Work: How to Reduce Risk When Nothing Goes Wrong

Paperback ISBN: 978-1-0681964-0-9
Hardback ISBN: 978-1-0681964-1-6

www.learningfromnormalwork.com

What Industry Leaders Are Saying

"This engaging guide on transforming workplace safety delivers a practical roadmap for organizations transitioning toward a learning culture. Ideal for safety leaders seeking culture change, reduced risk, and authentic engagement with frontline workers."
—Andresa Hernandes, Vice President of Safety, Siemens AG

"With practical examples and actionable advice, it offers useful strategies to build safer, more effective teams and improve performance."
—Gøran Furland, VP QHSE, Odfjell Drilling

"Learning From Normal Work is simple and powerful, taking you from concept to principle to practice in an entertaining and masterfully informative industrial novel. I was taking notes on almost every page and I will definitely be providing this book to my leadership and practitioner teams. This book is a goldmine of ideas and practical steps.... Read it, highlight it, learn from it,

and put it into practice. Learning From Normal Work will change the way you learn, you engage, and your results."
—**Charles Major,** Director of Operational Excellence and Human Performance, Luminant | President at the Community of Human and Organizational Learning

"A refreshingly easy and inspiring read on how learning from everyday work leads to meaningful change."
—**Donald Lyons,** Senior Director, Global Workplace EHS/OH Excellence, Kenvue

"What really sets this book apart is how clearly it shows the importance of listening. When workers trust that you won't weaponize their honesty, they start offering the kind of insight that actually prevents accidents."
—**Thomas Stewart,** Director of Safety & Security, Range Resources

"I would recommend this book to all safety practitioners who have tried traditional approaches with little or no success but desire to see real change...for the better!"
—**Keron Joseph,** Head, Process Safety Management, National Energy Corporation of Trinidad and Tobago Limited

Table of Contents

Acknowledgements

This book has been in development for a long time. It was inspired by the failures and successes I witnessed while introducing proactive learning in various organizations.

It took time to absorb the lessons from those experiences, then more time to learn how to write a novel, followed by many months of refinement.

I'm grateful to the many people who supported me and helped to shape this book. Thank you to Piotr Cichowicz, who offered his feedback and critique in the early versions of the draft, and to the many people who offered their time and effort to review the beta version of the manuscript: Mitchell Cini (ABB), Byron Ardoin (CSP), Jean-Pierre Fry (Domtar), DeWayne Bone (Domtar), Lamberto Nonno and Abdelasalam Yasseen (Baker Hughes), Gretchen Haskins (HeliOffshore), John Wilkes (Novavax), Diane Chadwick-Jones (ex-BP), Stuart Edmonston (UK P&I Club), Stuart King (Energy Institute), Cindy Nandlal (Point Lisas Nitrogen Limited), Andresa Hernandes (Siemens AG), Donald Lyons (Kenvue), Thomas Stewart (Range

Resources), Keron Joseph (National Energy Corporation of Trinidad and Tobago Limited), Gøran Furland (Odfjell Drilling), and Charles Major (Luminant).

Also, to my editorial team, and Chisom Ezeh of The Publishing Pad for your invaluable help in refining this book.

You have all helped to shape the story and define the lessons we can learn from workers on the front line.

Introduction

In my professional practice, most of our customers approach us after encountering concepts such as Human and Organizational Performance (HOP), proactive learning, or related principles. They often say, "*We like these ideas, our leaders are on board but what now? How do we apply these principles practically?*" The reality is that practical implementation is rarely straightforward. It is accompanied by numerous challenges, resistance, internal politics, and resource limitations.

Over many years, I've gained extensive experience implementing HOP and proactive learning programs, first within major global organizations such as BP and Baker Hughes, and later supporting other large corporations through my own company, Psychology Applied. Throughout this journey, I've witnessed a diverse range of situations, challenges, successes, and failures. However, openly discussing failures and specific organizational issues is often restricted due to confidentiality agreements and legal constraints. Moreover, challenges faced by one organization frequently differ significantly from those faced by others.

This reality inspired me to write this book. I wanted to create something genuine, practical, and engaging, rather than purely theoretical. The book's central character, John, is a senior Safety Director tasked with implementing Learning From Normal Work from scratch. John's persona was carefully constructed from the combined experiences, successes, and setbacks I've observed across multiple organizations. The dialogue throughout the book reflects authentic situations I've encountered, and most of the quotes come directly from real people dealing with real implementation challenges.

John's story isn't fictional in the traditional sense; instead, it synthesizes a wide variety of real-world experiences drawn from different companies, industries, and scenarios. Although certain elements have been simplified for clarity and storytelling purposes, every situation John faces closely mirrors the complexities, politics, practical constraints, and mindset shifts that genuinely occur during the implementation of Learning From Normal Work as a program.

My intention is to offer a practical road map that sheds light on what to expect when transitioning a large organization toward proactive learning, and to showcase the new mindsets, toolsets, and skillsets needed to do so.

Chapter 1

A solid safety record doesn't mean your facility is safe: the rude awakening

John prided himself on being a man of routine. Every morning, the loud beeping of his cell phone alarm woke him at exactly six-fifteen. He always had time to have breakfast with his wife and daughter and still arrive promptly at work.

Today, however, the cell phone woke him not at dawn, but at the unpleasant hour of four o'clock in the morning. The voice on the other end belonged to Mark, the Plant Manager at Alpha Facility, his voice tight with urgency.

"John, we've had a serious accident. One of the crew is injured, and it's bad. We need you here, now."

John threw off the covers, adrenaline surging through his veins. Accidents were a rarity for Avriox Industries. John's Division had just celebrated two years without a single lost-time injury, so this was a brutal wake-up call. John had spent the previous evening at a celebration dinner with his colleagues, and his award for overseeing the two-year perfect health and safety record as Division Director was still on the kitchen table. It felt meaningless now.

John drove to the plant, a tense knot forming in his stomach. The atmosphere on site was buzzing with a mix of disbelief and fear. John was immediately briefed on the situation. The accident had happened during a routine operation to move a large metal spool. The task required the operator to lift the spool half a meter or twenty inches above the ground and transport it across the hall using a crane. During this process, the spool had struck one of the workers, a young man named Harry.

Harry had been hired a year ago and was known for his dedication and hard work. He had offered to help a colleague, and keep his supervisors happy, by catching up with work on a project that was running late. This led to his presence at the workshop overnight, welding additional components. Wearing a welding mask, he couldn't see or hear the approaching load. Unaware of the incoming danger, he was hit by the metal spool, which weighed over eight tons.

By the time John arrived, an ambulance had taken Harry to the hospital with serious injuries. While his life was not in danger, his recovery was likely to take several months.

What struck John most was that the crane operator was Dan, a veteran employee with over twenty-five years of experience.

A paramedic was still checking Dan when John walked into the workshop. He pulled John aside and warned him that the accident would probably affect Dan's immediate and long-term mental health. The operator was in shock after seeing what had happened to Harry.

"He might find this hard," the paramedic explained. "Extreme guilt, high levels of stress. Please consider offering him mental health support and assigning him a different task for the next few weeks."

John realized that this accident would have a profound impact on Dan, despite his extensive knowledge and skill.

John was trained in Root Cause Analysis. He joined the Alpha Site investigations team as they put up barriers around the crane and the accident site and began their evidence collection. Photos, diagrams and witness interviews would all be useful as they investigated the cause of the accident.

Avriox Industries, like many companies, utilized the 'Five Whys' investigation method. Every time the team discovered a potential cause for the accident, they asked why that cause existed. After asking 'why?' five times, it was assumed that they would uncover the root cause.

Days later, after multiple interviews and careful analysis, John's team was ready to present the findings to the anxious company management team. The culprit, according to the initial investigation, was human error. The crane operator, it seemed, had made a critical mistake in not noticing where Harry was working.

"But why did it happen?" the management team asked.

"Complacency," John concluded, the weight of the word hanging heavy in the air. "We will retrain all crane operators, run a communication campaign about our zero accidents mission and the importance of compliance, introduce a spotter to ensure the path is clear, and ask every single person to sign a pledge to always follow all the rules and pay attention to what they are doing," he said.

A satisfied murmur rippled through the room.

Following the accident, blame was officially placed on the crane operator for his lack of caution. He was given a final written warning and required to undergo retraining to learn to be more vigilant. In addition, he faced an uncomfortable conversation with his supervisor about the importance of paying attention.

As Division Director, John also felt pressure to improve his own performance. He felt as if he hadn't deserved his award for health and safety management, and he still felt uneasy that the blame for the accident had fallen on Dan, in spite of his good safety record and years of experience.

Exhausted from stress and long hours, John found himself sitting at his kitchen table after dinner, running through the accident investigation in his head. His wife Melissa took the seat beside him.

"What's bothering you?" she asked, concerned. "Is it the accident? You've been like this for days ... lost in thought, not really here. What's going on?"

John shook his head. "I'm not sure. I can't help thinking that there's more to the Alpha Plant accident than one mistake from a single person. It seems more complicated than that."

Melissa nodded. "You did say that the crane operator was good at his job, and that the injured man wasn't to blame. So, what else could it be?"

"I wish I knew," John said. "We followed all the right procedures to uncover the cause, but 'human error' just feels incomplete. I think we can do better, but I don't know where to start."

The thought made him anxious. How could he find answers when there was so much he didn't understand?

Chapter 2

You can spot (and prevent!) accidents before they happen: making the invisible, visible

The accident cast a long shadow over John. Images of the hanging load and the empty space on the crew roster haunted him. The injured worker remained in the hospital.

John rubbed his temples and sighed deeply. "What could I have done to prevent this?" he muttered to himself. "What did we miss?"

The risk assessment for the crane operation had been his responsibility. He had signed off on it, confident that all safety measures were in place. Yet, here they were, dealing with a tragedy.

His guilt over accepting the safety award just before the accident was echoed in his conversations with the management team. He felt certain that they assumed the award had been a mistake, and he wondered what the consequences might be for his career.

John's thoughts drifted to Harry's family, his wife and kids. The pain they must be going through was unimaginable. He felt the heavy burden of guilt pressing down on him. His two-year streak of zero accidents felt like a hollow achievement.

He spent nights scouring the internet, searching for a different way to approach safety management. As he read, an interesting concept caught his eye: Learning From Normal Work. The idea was simple, yet profound. Understanding exactly how workers completed their daily tasks, regardless of the 'correct' procedures, could allow him to identify risky activities before they turned into accidents.

He read various articles and case studies, each one explaining more about this proactive approach. The reactive techniques John was trained in seemed so different, mainly learning from accidents after they happened. John wondered whether Learning From Normal Work could help prevent accidents like the one that injured Harry and give a better insight into the causes of unsafe behavior.

The next day, John had a scheduled lunch with Karen, the Vice President of Operations. They met at a quiet restaurant near the office. As they settled into their booth, Karen looked at John with concern.

"How are you holding up, John?" she asked.

"It's been tough," John admitted. "I can't stop thinking about Harry and his family. I keep wondering if there was something more we could have done to prevent the accident."

Karen nodded sympathetically.

"I've been thinking a lot about it too. In fact, I wanted to discuss something with you. At the supplier forum last week, I learned about a concept called Human and Organizational Performance, they called it HOP. There was a presentation on how to implement HOP, focusing on Learning From Normal Work (LFNW). Have you heard of it?"

John's eyes widened in surprise. "Actually, yes. I was reading about it just last night. It seems to be about understanding day-to-day operations. Finding out what workers actually do and identifying problems we don't know about because we don't do that work. A way to spot the adaptations on the shop floor before they cause accidents."

Karen looked intrigued. "Exactly. The presentation was really interesting. It made me think that we might be missing something in our current approach. What if we could learn from the real, everyday activities of our workers, and use that knowledge to prevent accidents?"

John felt a spark of hope. "I think we should explore this further. Could we invite an expert to give us a presentation like the one you went to? That could help us understand how to use these concepts in our official safety protocols."

Karen smiled. "I'm glad we're on the same page. Let's set it up."

••●••

A week later, John organized an internal presentation with a HOP expert. The conference room was filled with the company's senior leadership team.

The expert began her presentation by introducing the core principles of Learning From Normal Work.

"'Normal Work' is about how people adapt to changing conditions and challenges as part of their job. For example, every time a crane operator uses the crane to lift a load, there may be something different about the situation. They might have less time available than planned, there may be additional people in the area, or perhaps the right type of lifting sling is not available. Each time they use the crane, they need to adapt to the challenges around them. This is Normal Work, and these challenges that workers face are called 'constraints' or 'error traps'.

"An 'error trap' is anything that makes it more likely that a person will make mistakes, or not follow safety rules, when carrying out their work. A 'constraint' is something that limits, or constrains, the choices available in a given situation.

"Learning From Normal Work involves listening to the people who do the work about the challenges, error traps, and frustrations they face. Many of these problems and error traps are related to resource limitations, and how jobs are set up, and many are not directly observable," she explained. "It's about recognizing the problem-solving that workers do to get the job done. These adaptations can reveal underlying risks that aren't apparent in formal risk assessments. It's about making the invisible, visible."

John watched as the expert continued her explanation, and the team absorbed the information.

"Proactive learning means shifting our focus from investigating accidents after they happen, to understanding and improving Normal Work processes. We need to understand what the people who do the work *actually* do, as opposed to what they are *expected* or *trained* to do. This approach can help us identify and reduce risks before they turn into accidents."

The room buzzed with information, Human Performance, error traps, adaptations, all factors that were influencing behavior on company sites.

John felt a surge of optimism. Learning From Normal Work wasn't about finger-pointing after accidents. It was about proactive learning and discovering the invisible adaptations that increase risk in the workplace.

As the expert concluded her presentation, a hush fell over the room. John, eager to capitalize on the management team's reaction, asked, "What do you think? Could this be something we implement at Avriox Industries?"

A frown creased the Director of Operations' brow. "But John," he said, "we are already doing this. We already have a behavioral safety program. We identify unsafe acts. We conduct risk assessments. Isn't this just another layer of bureaucracy?"

John's heart sank.

Another manager chimed in. "It sounds interesting, but is it really worth our time?"

John was disappointed that Karen, who was also in the room, stayed silent. The room remained divided.

Two days after the meeting, Karen called John, asking to pause the Learning From Normal Work idea for now.

John, frustrated but determined, knew that the management team's buy-in was essential. Winning them over had failed. The concept of Learning From Normal Work had intrigued them, but it challenged their longstanding established ideas, safety practices and processes.

Reluctantly, John abandoned the idea of implementing a proactive approach to health and safety, but he remained curious. He continued his research into HOP and Learning From Normal Work in his free time, confident that this approach could lead to better health and safety outcomes for the company, if he could only convince the management team.

A few days later, John was in a meeting, reviewing the latest safety reports, when his phone buzzed on the table. He glanced at the screen and felt his stomach drop, a call from the Gamma Facility plant manager.

"John, we've had another accident."

Chapter 3

Deja vu, with a difference

The manager's voice was strained with controlled fear. "Same activity as the Alpha Facility incident," he said. "The worker's injured, but not as seriously this time. It's a miracle she wasn't crushed."

John felt a cold sweat break out on his forehead. "I'm on my way," he said, grabbing his jacket and car keys and heading for the door.

Gamma Facility had stopped all production by the time he arrived. Paramedics were tending to the injured worker, who was conscious but clearly in pain. The scene was all too familiar, and it filled John with a mix of fear, frustration, and helplessness.

The crane operator, a relatively new employee named Tom, had been tasked with moving a large metal spool across the hall. After the last accident, spotters had been appointed to help with crane operations. Today's spotter, Lisa, had been equipped with a radio to communicate with Tom. Despite their efforts to coordinate, the tight space in the hall made it difficult to move the spool.

As Tom lifted the spool, it had started swinging unpredictably. Unfortunately, despite Lisa's efforts to stay clear, the spool swung into her path, striking her as she tried to move away. Unlike Harry, Lisa managed to partially shield herself at the last moment, resulting in less serious injuries.

John approached the plant manager, who looked equally shaken.

"How did this happen again?" John asked, his voice barely concealing his anxiety mixed with anger.

"We tightened up all the procedures after the last accident," the manager replied. "We even had everyone sign pledges to be more careful. I don't understand it."

John sighed deeply. "Let's secure the site and start the investigation. We need to figure out what went wrong."

The atmosphere on site was tense. Workers were worried that management might decide to punish anyone they found responsible, especially as the safety procedures had so recently been reviewed.

As John and his team began their investigation, his phone buzzed again. This time, it was the CEO. John's heart pounded as he answered.

"What's going on, John?" the CEO's voice was angry and accusatory. "I thought your division was safe. You've had the best record for safety, and now we've had two serious accidents in a short period. Who screwed up? Who should I fire? Is it you, John?"

The CEO's words came as a shock, adding to John's anxiety. He felt concern, fear, and a profound sense of humility. He understood the CEO's frustration, but the aggressive tone wasn't helping anyone. It certainly wasn't helping his attempts to learn from these accidents.

"We're investigating the accident now," John replied, keeping his voice steady despite his nerves. "We'll find out what happened and take the necessary steps to prevent it from happening again."

"You'd better," the CEO snapped. "I want answers, and I want them fast."

John hung up, feeling the weight of the investigation on his shoulders. He knew they had done their best to address the safety issues after the first accident. They had retrained all their crane operators, launched a communications campaign, reminded all operators to be vigilant, and required everyone to sign pledges to follow all the safety rules. Yet, despite all these efforts, they were facing another serious accident.

As the investigation progressed, John couldn't shake the feeling of déjà vu. They used the 'Five Whys' method and found that, once again, human error and complacency appeared to be the most likely root causes. The injured worker and the crane operator had deviated from safety protocols, just as they had for the first accident. The training and reminders hadn't been enough.

John gathered his Division Safety Team in the conference room. The mood was tense. "We need to understand why our efforts aren't making a difference," he said. "We've done everything by the book, but it's clearly not enough."

One of the supervisors spoke up, frustration evident in his voice. "People got retrained, but they're still making mistakes. How do we fix that? We need stronger accountability!"

Meena, a member of the safety team, chimed in. "Maybe we're missing something fundamental. We need to look deeper into how our workers are actually performing their tasks."

John nodded, his mind racing. He decided to share his research into Learning From Normal Work. "We need to move beyond just telling people to be careful," John said. "We need to understand the complexities of their work and find ways to support them better."

The team listened intently. They had faced another blow, but John wondered whether this could be a turning point. He could see that they needed to develop a deeper understanding of the work processes on site and consider how to make them safer.

Leaving the conference room, John felt a renewed determination to change the company's approach to safety and accident prevention.

••●••

John was sitting in his office, feeling the weight of the recent accidents in his division. The Root Cause Analysis reports sat on his desk, but they felt incomplete. Two similar accidents, on different sites in the business. John couldn't shake the feeling that there was more to the story.

Desperate for a fresh perspective, he reached out to Sarah, an old friend who worked at a company known for its advanced safety practices.

"Hey, Sarah. It's John. Do you have a moment to talk?" John asked over the phone, his voice tinged with frustration.

"Of course. What's going on?" Sarah replied, sensing the urgency in his tone.

John explained the two accidents, detailing the injuries and the conclusions from the investigations. "We identified the root causes, and we made some changes, but I can't help thinking we're missing something. The solutions seem too narrow. It feels as if we're only addressing the symptoms, not the underlying issues."

Sarah listened carefully. "It sounds to me as if you're hitting the limitations of traditional Root Cause Analysis," she said. "Can I share an example from my site? It might help explain what I mean."

"Of course," John responded, happy to listen to Sarah's experiences.

"I'm emailing you an image," Sarah said. "Take a look, and I'll explain."

"One day, at our industrial site, a worker twisted his ankle while disposing of a bag of waste into an industrial bin," Sarah began. "These bins are more than two meters or six feet high, and to throw the bag inside, the worker had to climb the side of the container, as you can see. Unfortunately, his foot slipped, and he landed awkwardly on his ankle.

"Our first response was to conduct a Root Cause Analysis," Sarah continued. "We relied on a popular Root Cause Analysis methodology, and we even took a refresher training course before we started the investigation.

"I remember the trainer emphasizing that things that did not contribute to the event should not be added to the cause-and-effect diagram.

"So, we identified that the container didn't have proper steps, just a support beam that the worker climbed on. His foot slipped from the beam, causing the injury. The corrective action was to modify the bin design to include proper steps."

"That seems straightforward," John said. "But I'm guessing it didn't solve the problem."

"No," Sarah replied. "Turns out it was a classic case of addressing the immediate cause without considering the wider context. The modifications probably prevented a repeat of that specific accident on that specific bin, but didn't account for other variables.

"We realized that the situation was more complicated than problems with this one bin, and one single solution. We decided to try something different, and use a Learning

Team approach," Sarah explained. "Instead of asking why the accident happened, we went to the workers, and simply said, 'Tell us about what makes this work difficult.' This shifted our focus from single causes to understanding all the challenges the workers faced when using the bins."

John was captivated. "So, what did you find?"

"Some really interesting results," Sarah said. "Through these discussions with the workers, we discovered several critical pieces of information.

"The site used industrial bins from four different suppliers, and there were ten different models of bins in use on the site. Four out of the ten bin models did not have proper support structures for safe climbing.

"We also realized that not all of our workers were physically able to climb some of the bins. They could only use one hand to climb, as the other hand would be holding the bag.

"We came up with an initial solution that would work for all the workers, and all the bins."

"Wow," John said, surprised. "So, the problem was much bigger than just one faulty bin."

"Exactly," Sarah confirmed. "The Learning Team approach showed us that the issue was wider than we had initially understood. So eventually, instead of just replacing this one bin model, which would have been a box-ticking exercise, not a solution, we worked with our suppliers to standardize the type of bins used at the site. We developed a set of requirements and made sure that every bin had proper steps or support structures.

"We involved the workers in the process, using their experiences to bring in genuinely safer practices and equipment. We reached out to suppliers and made sure that the new bins complied with these new standards. Instead of ticking boxes and looking as if we had fixed the issue, we dug deeper and actually solved the problem."

John was frantically taking notes.

1. Root Cause Analysis identified the specific design flaw with the bin involved in the accident, but **did not** consider the wider context of bin use across the site.

2. It failed to account for how different bins with different designs affected the risk. Root Cause Analysis did **not** show that multiple bin models posed similar challenges.

3. Corrective actions from the Root Cause Analysis were too specific to the bin involved in the accident. This would **not** prevent similar accidents involving other bin models without proper support structures.

4. By addressing **only** the root cause, they risked creating a false sense of security. Leaders might think the issue had been resolved, while similar hazards remained.

John sat at his desk, staring at the phone after his enlightening conversation with Sarah. He couldn't shake the feeling that he had missed something important in his own investigations. He glanced at the reports from the recent lifting accidents, feeling a growing sense of unease.

"Sarah's right," John muttered to himself. "We've been focusing too narrowly. We've been settling for simple fixes without understanding the full context."

He picked up the report from the first lifting accident. The conclusion was simple: the worker had not followed the procedure, and the corrective action was to reinforce training on those work instructions.

John could see that, like the bins in Sarah's story, these solutions were fixing the symptoms, not the underlying issues. Determined to dig deeper, he decided to approach the problem from a new angle, inspired by Sarah's Learning Teams.

He realized he was starting to see accidents, not as one-off mistakes, but as symptoms of deeper issues. He worried that disciplinary action and retraining of the workers involved wouldn't be enough to fix the real problems.

He also realized that if he wanted to keep his employees safe, he couldn't wait for the next accident to happen before looking for a solution.

John's phone vibrated in his pocket, a call from the Vice President of Human Resources. She asked John to attend a meeting in three days to discuss his recent performance.

"Am I going to be fired?" John asked himself, his heart racing.

Chapter 4

The blame game: the problem with fault-finding and scapegoating

John was distracted as the meeting with HR drew closer. He didn't know what to expect, and on the day of the meeting, he found himself awake before his alarm clock sounded. He crept out of bed and headed into work early.

"John. Come in. Take a seat." The Vice President of HR held the door open for John as he stepped into her office.

"I'll make this quick," she said, sitting down behind her desk. "You're here because your division has reported two serious accidents in a short period of time. The prevention of accidents is a major part of your job description. With your recent safety record, we have concluded that you are not meeting your contractual obligations to the company."

John's heart was racing. He felt sure he was about to lose his job.

"Today I am giving you an informal verbal warning," the VP continued. "I need to inform you that further action will be taken if the safety standards in your division continue to fall below expectations."

The rest of the meeting passed quickly. John could only focus on his relief at keeping his job, and his determination to change the company's approach to safety, investigations, and blame.

•●•

A few days later, John spent the day on the factory floor at Alpha Facility, where Harry's accident had happened, talking to operators. The fear and distrust among the workers he spoke to was obvious, and after his meeting with HR, he could understand their concerns. His worries grew stronger after a casual chat with another operator, Simon. Despite his forced smile, Simon's eyes were filled with worry.

"Morale's terrible, John," Simon admitted. "Since the accident, everyone's scared and angry. We all knew about the problems that led to Harry's injury, but management just doesn't listen. And to be honest, your investigation was awful. The guys are upset. Some want to quit, but they're stuck because of the pension plan. Others are talking about slowing down work on purpose."

Simon's anger was clear, but John didn't take it personally. He sensed that the operator's frustration was rooted in a deep disappointment with the management team.

Simon had hoped for a stronger focus on safety as a result of John's investigation, but instead the result had spread fear among the workers. John could see that this wasn't just a morale issue. Productivity had dropped, and management had noticed.

One veteran operator named Karl summed up his frustration. "We're damned if we do, and damned if we don't. Follow the procedures to the letter? Production quotas suffer, and we're expected to do better. Find a workaround to get the job done? The blame game starts."

••●••

It wasn't a surprise when the CEO summoned John and Karen to his office. The air crackled with tension.

"What the hell is going on, guys?" the CEO boomed. "Production is down, customer deadlines have been missed. What are you people doing?"

John tried to explain the impact of the accidents, the results of punishing the workers with disciplinary measures, and the fear that had gripped the workforce.

His words were met with skepticism. "Excuses! We need results!" the CEO scoffed. "I thought HR had made that clear."

John felt increasingly isolated. The blame game had taken hold, pitting departments and divisions against each other, each scrambling to avoid responsibility.

Discouragement gnawed at John. The frustrated workers he'd spoken to weren't casual employees who could be easily replaced, they were the lifeblood of the plant. Their experience and knowledge were invaluable. But management seemed blind to the fear that was driving them away.

Upstairs, the finger-pointing continued. After the meeting with the CEO, John overheard the executive team discussing disciplinary measures for him if his record didn't improve. Their focus was solely on punishment, a knee-jerk reaction with no consideration for the bigger picture.

John felt trapped. The blame game was fueling fear, eroding trust, and ultimately impacting performance. He knew he had to act, but how could he break the cycle and convince the company to embrace a new approach to safety? They needed to start focusing on learning over punishment.

He spent the next few days speaking with more operators. The stories he heard painted a grim picture. Workers were demoralized, fearful, and distrustful of management. Comments like 'should I lie to the bosses to save my job?'

and 'management is more interested in ticking boxes than listening to us', were alarmingly common.

He realized the blame game had deeper roots than he had initially thought. The broader management team, including the CEO and VPs, were caught up in their own finger-pointing. Heated discussions about who should be fired were commonplace. The CEO's earlier anger had only fueled this environment.

In one particularly tense meeting, the CEO's voice boomed across the boardroom. "We can't let this poor performance continue! Someone has to be held accountable. If people are breaking the rules, they deserve disciplinary action. It's that simple."

The VPs nodded in agreement, each eager to shift the blame away from their own departments. "We need to set an example," one VP said. "Reprimand, suspension, firing, whatever it takes to make it clear that rule-breaking won't be tolerated."

John sat silently, a knot of frustration and helplessness tightening in his chest. The conversations revolved solely around punishment and consequences. There was no discussion of underlying issues or proactive solutions. He knew that this approach was not only unfair, but also counterproductive.

After the meeting, John returned to his office, his mind a whirl of conflicting thoughts. "If even the long-term employees are thinking about leaving, should I?" he wondered aloud. "HR clearly thinks I'm not doing my job. Maybe it's time to move on."

With two accidents and a verbal warning on his record, his chances of promotion were bleak. The culture of blame and fear was stifling any hope of real improvement.

But John was determined not to give up. He decided to take a different approach. He gathered his notes and prepared a report that highlighted the insights he had gained from his conversations with the operators. His goal was to show that the issues went beyond individual mistakes, they were rooted in widespread problems that required comprehensive solutions.

•••••

The next day, John requested a private meeting with the CEO. He entered the CEO's office with a mix of apprehension and resolve. The CEO looked up from his desk, his expression stern.

"John, what is it now? More bad news?" he asked curtly.

John took a deep breath. “I’ve been speaking with the operators, and I’ve learned some troubling things.

“Punishing individuals isn’t solving the problems, it’s making them worse. Morale is low, productivity is down, and there’s a sense of fear among the workers. We need to shift our focus from blame to understanding the real challenges they face every day.”

The CEO’s eyes narrowed. “So, what are you suggesting? We just let people off the hook?”

“Not at all,” John replied, maintaining his calm. “I’m suggesting we look deeper than the behavior and try to understand whether we are setting our people up for failure.”

The CEO leaned back, considering John’s words. “Alright, John. I’ll give you some leeway to explore this further. Run a small pilot and bring me practical examples. I want to see real value and insights from you.”

John nodded, a mix of relief and determination washing over him.

He needed answers. The future of the plant, and his career, depended on it.

Chapter 5

John struggles with proactive learning

Inspired by Sarah's advice, and the CEO's challenge, John paid another visit to Alpha Facility. He had been reading the International Association of Oil and Gas Producers (IOGP) industry guidance on Learning From Normal Work, and he wanted to put some of his insights into practice. He joined two crane operators in the canteen during lunch, hoping to gain insights into their daily challenges.

"Alright, guys," John began, trying to sound confident. He jumped straight to the point. "I want to understand what makes your work difficult, especially when lifting heavy objects. Why do you sometimes cut corners or ignore hazards?"

The workers seemed offended, exchanging uneasy glances. After an uncomfortable silence, one of them, Andy, spoke up. "Well, we try our best, but sometimes we have to skip some steps in the process, otherwise the work would take too long."

John frowned. "But what about following procedures and staying safe?"

They fell silent and seemed reluctant to speak further. After a few more awkward exchanges, John left the conversation without picking up any useful insights.

Back in his office, John cursed loudly. "Damn it! If I don't make this work, I've got nothing else to offer. We'll be using more of the same old safety programs forever."

He knew he needed help, and there was only one person he could think of. John picked up the phone and called Sarah.

"Sarah, it's John. I need your help."

"What's up, John?" Sarah replied.

John sighed. "I did as you suggested and tried to review a lifting activity with the workers, but it didn't go well. I don't know what to do, and I don't have much time left to convince the CEO that we need to change our approach to safety."

Sarah listened patiently and then invited John to visit her factory to observe her work.

When John arrived at Sarah's site, they sat down for a chat before heading to the shop floor.

"So," Sarah began. "Tell me about your experience, and how things went."

John wasn't expecting challenging questions from Sarah, but he was keen to learn from her experience. "Well," he said, trying to decide what to tell her. "I joined them at lunch and asked them questions like, 'Why did you do *this*?', 'Have you followed all the procedures?', 'What are the hazards?'

Sarah nodded thoughtfully. "And how far did you get with those questions?"

John shook his head. "Not very far. They didn't open up much. It was awkward."

Sarah smiled. "Those questions sound good, but they aren't effective for Learning From Normal Work. You need to ask non-leading, open questions, and give people the chance to explain their concerns. When we go to the shop floor, I'll show you the TEDS technique. It stands for 'Tell, Explain, Describe, and Show', and it really helps people to open up."

On the shop floor, Sarah gathered three workers for a chat. She made them feel at ease, addressing them by name and asking about their families and children.

Sarah began warmly. "Help me to understand the key challenges you face with this lift you are working on today."

One worker, Rasheed, started to describe their process.

Sarah listened attentively, using TEDS to dig deeper. "Tell me more ... Describe a situation when it was particularly tough ... Show me what you mean."

In just fifteen minutes, Sarah had gathered a wealth of insights. John watched in awe, realizing how much more effective this approach could be.

On his way home, John practiced asking TEDS questions in his mind. Open questions and give people the chance to respond. When he got home, he tried it with his family.

"Could you tell me why you didn't do your homework?" he asked his daughter, Erica.

She gave him an angry look and stormed out of the room. Undeterred, he tried again with his wife.

"Tell me about your day," he said. Melissa looked surprised. She could not resist the invitation and started describing her day in detail. John interrupted her after a few minutes with a judgmental, "Really? Why on earth did you do that?"

Melissa paused, clearly annoyed. She took a calming breath and said, "John, it was lovely to hear you ask that question. Normally you say 'I hope you had a good day' and move on, but this time you gave me a chance to tell you about it, and it felt ... wonderful. But challenging me while I'm talking? You ruined it." She smiled. "Sometimes you need to just shut up and listen."

John was shocked to hear the frustration in Melissa's voice. She had seemed so happy to tell him about her day, but when he interrupted, she reacted with anger.

He felt confused. "So ... I shouldn't ask more questions?"

Melissa put a hand on his shoulder. "Questions are good, but you need to listen to the answers. You need to really hear what I'm saying. Give me space to talk, before you make it all about you again."

John spent the evening trying to understand where he had gone wrong. He'd asked an open question, and he had listened to begin with, but by challenging the answer, he had upset Melissa. He thought about how Sarah spoke to the workers on her site, always listening, and giving them simple prompts to make sure they shared all the information she needed.

Returning to Alpha Facility the next day, John decided to try again. This time, he prepared his questions in advance and focused on a specific type of lift. He asked the supervisor to give three workers a thirty-minute break and gathered them in the work area, trying to replicate what he had seen at Sarah's site.

This time, he started differently. He introduced himself and explained the context and purpose of the conversation.

He learned their names and asked about their day. He reassured them about the confidentiality of the conversation.

"Can you tell me about the challenges you face with this particular lift?" he asked, using the TEDS technique.

The workers responded well, sharing a few useful insights. However, there wasn't an avalanche of problems. They were clearly checking John out and testing to see what he would do with the information. It felt like a very controlled drip feed. The workers didn't want to bluntly refuse to speak with the director, but they also didn't trust him enough to open up.

One worker mentioned the need for a different model of push sticks. John acted quickly and agreed with the site manager to order them the same day. When they arrived the following week, the workers were surprised they had been taken seriously but remained suspicious.

John decided to spend more time chatting with the workers, focusing on building trust. He spent two hours with four workers, using TEDS to look at another type of crane lifting activity.

Gradually, and after about an hour of trust-building exchanges, the workers began to open up more, sharing insights about conflicting priorities and unrealistic schedules. Still, John didn't feel he had reached the full potential of this approach, although he was clearly making progress.

•••

The following week, after reading more of the IOGP guidance, he decided to try a Walk-Through Talk-Through, a simple technique that involves going through the steps of a task individually and discussing what made each step difficult. John learned that different steps were affected by unexpected error traps and could fail in ways that would be missed if he focused on the job as a whole.

John used a simple form to help him structure the process and made notes on everything he discovered during his walk-through talk-through conversations.

Step	**Error Traps**	**Improvements**
Preparing lifting equipment and choosing correct lifting slings	Selecting wrong slings	Always ensure the correct tool is selected before operating the equipment
Rigging the equipment (attaching the spool to the crane)	Not paying attention.	Remind operators to pay attention.

Step	Error Traps	Improvements
Moving the spool with a crane	Using hands to push the sus-pended load.	Remind operators never to touch the spool with their hands.
Setting the spool on the ground and securing to prevent further movement	Overconfidence and underesti-mating the risk.	Remember to not be overconfident, and do not under-estimate the risk.

Back in his office, John decided to send his findings to Sarah for feedback. He typed up an email detailing the error traps and improvements he had identified and hit send. The next morning, Sarah responded, suggesting a quick call to discuss his walk-through.

On the call, Sarah started with a gentle tone. "John, I see you've put a lot of effort into this, and that's great. But I think there are some areas where we can improve."

John could feel his disappointment at Sarah's words, but he encouraged her to continue.

"The error traps you've listed, lack of awareness, not pay-ing attention, selecting the wrong tool, these are not actual error traps. They're more like behaviors," Sarah explained. "This is a common mistake for people doing a walk-through talk-through for the first time. Let's break it down."

John took out his notes, ready to absorb Sarah's insights.

"Instead of saying 'lack of awareness', think about what process might lead to that lack of awareness. For example, does the training cover everything the workers need to know?" Sarah suggested.

John jotted down her points.

"And for 'not paying attention', ask yourself what might be competing for their attention. Is there someone asking them questions while they work because they need help?"

"Selecting the wrong tool," Sarah continued, "is a behavior. Instead, look at what factors increase the chances of choosing the wrong tool. Is the tool storage organized? Are the tools clearly labelled?"

John's pen moved quickly across the paper.

"Using hands to push the suspended load is certainly an unsafe behavior, but why do they put their hands there? Maybe it gives them better control over the load."

John felt a light bulb snap on in his head. "So, I should focus on understanding *why* they do things a certain way, not just *what* they're doing wrong."

"Exactly," Sarah said. "And phrases like 'underestimating risk' and 'overconfidence' are judgmental. They don't help us understand the work. Instead, focus on what makes the work difficult, and how the process can be improved."

The next day, John tried again, and this time he felt more confident. His new walk-through talk-through table looked much more useful.

Steps	**Error Traps**	**Improvements**
Preparing lifting equipment	Slings look the same. Hard to choose the correct safe working load.	Add tags with different colors to make it easier to choose the correct sling.
Rigging the equipment	Rigging training was a long time ago, and operators don't remember rarely used lifting techniques.	Schedule refresher training for crane operators and riggers on annual basis.

Steps	Error Traps	Improvements
Moving the spool with a crane	The spool is 2.5m in diameter and it's hard to see what's behind. The wired control panel is restricting movement and visibility.	Provide remote controls, so the crane operators can move freely.
Setting the spool on the ground	The spool is round, so it can easily move once it is lowered. It's dangerous to step close to put down wedges.	Install permanent wedges on the floor, 15cm above the ground, so they can put the spool behind and not worry about movement.

John reviewed the table. He noticed that dangerous behaviors, like pushing the spool during lifting, would usually result in disciplinary action. But after his walk-through talk-throughs, he could see that these actions were simply adaptations to get around the error traps of the job. Rather than punishing the workers, he needed to understand these error traps and offer improvements and solutions instead of discipline.

When he shared his second table with Sarah via email, she responded with a big thumbs up and a smiley face.

John spent time practicing over the next week. When he had gathered enough insight into the process of using the crane, he felt confident to take the results back to his team.

••●••

"I can't believe how we set them up for failure!" exclaimed John, as he walked into the conference room to attend the site management meeting. He was a few minutes late, and his peers were already inside waiting for him.

The meeting participants looked at John with surprise and confusion.

"What are you talking about?" asked Grace, the Site Manager, sounding irritated that John had not provided any context.

"Do you remember the presentation from the Learning From Normal Work expert a few months ago? She told us that we could identify the precursors of the next accident, without waiting for the accident to happen. We cannot have a third accident in this Division, so I decided to give it a go."

"And...?" asked Grace.

"It wasn't easy, and I needed help, but I learnt a lot." John took a breath and began to outline his methods and his findings.

"At first, the operators were a bit hesitant to speak up," he explained. "They were worried they might get in trouble if they pointed out problems in their workspaces, and they weren't sure how their suggestions would be received.

"But eventually they started sharing their experiences. They told me about the confusing design of the controls for the cranes, and how they often had to stand too close to the load because of the unreliable signal from the remote control. They shared other issues, like problems with the maintenance of the cranes, and the lack of clear instructions for certain tasks."

The other managers were initially skeptical and surprised that John would spend time on something that didn't result in resolving a real accident.

"There will always be some minor issues, John," one of them pointed out. "And we need to concentrate on the accidents we do have. We give people tools and procedures, and we expect them to use them. Why change now?"

"Bear with me," John replied. "That's exactly what I thought to begin with. But I can't believe all the things I've found out! Despite all the investigations, audits, and procedures we put in place, we've missed so many issues. If I had known about these problems earlier, we could have fixed them and prevented the accidents."

"You know we have our life-saving rules, right?" asked Gary, the Engineering Manager.

"And you know that for a while we've been finding crane remotes that have been opened, and the electronics messed with?" replied John. "We tried to find and discipline the operators involved, with no luck. But when I talked to the operators, and asked them about the issues with the cranes, they were happy to explain. They showed me that our cranes are so old, the buttons on the remotes get stuck. When this happens, there is no way to finish a lifting job without taking the cover off and making connections with the wiring inside. They told me it's standard practice, and I'm sure we've only found the tip of the iceberg. This must be happening all over our sites," said John, excited to show his findings.

"But if workers have these problems, they should stop the job and tell the supervisor," exclaimed a visibly angered Procurement Director. "There is no excuse for this blatant violation of our safety rules. These rules are there to protect them and us."

"Yes," said John, "and we keep reminding them about consequences for breaking the rules, but we never actually listen to *why* the operators feel they *need* to break the rules.

"It turns out that the supervisors knew about the problem, but when they raised it with management, they were politely dismissed because there was no budget for repairs or upgrades. But when they stopped the job because it was unsafe, their monthly bonuses were reduced for missing the production quota, and they were told to find a way to solve the problem." John shrugged. "So they did."

John continued. "Did you know that if the remote control runs out of power during the lift, that stops the crane, and leaves the operators to deal with a heavy load swinging uncontrollably in their workspace? On a few occasions we've been really close to dealing with another high-impact collision. And there's no way to tell how much power is left in the remote. When the battery is low, it loses signal."

"But every crane has two remotes," said Ann, a new HSE advisor on John's team. "So, what's the problem?"

"True, but we haven't specified a standard storage location for the spare remote. Every team puts the second remote in a different place, which means that it's not always readily available when needed. That causes delays, and sometimes they can't find it at all.

"Not only that, but the labels on the remotes have worn through, making it difficult to read the controls and properly operate the crane. Phil, one of the operators, told me how one of our apprentices wanted the crane to move left, but because of the worn-out labels, he pressed the wrong button and moved the crane forward. Luckily, nothing was in the way."

John looked around at his team. Everyone seemed to be thinking about his findings so far. He hoped he was convincing them of the value of his research.

"We're setting them up for failure, you see?" he said. "The systems we've put in place don't reflect the reality of operating the cranes and keeping people safe. We discourage workers from reporting problems, we reduce their bonuses if they take longer with their tasks, and we discipline anyone who finds a way to fix the issues without reporting them."

Grace nodded. "I'm starting to see your point, John," she admitted.

"This is just the beginning," he replied. "For our next meeting, I'll prepare a full report."

Chapter 6

Finding the "causes" of accidents without waiting for an accident

On his way home, exhausted but inspired by his new insights, John decided to arrange a visit to another of the company's sites. He knew that Delta Facility had been accident-free for a while, and he wondered whether the issues he had reported to his team would also be found there.

The following week he arrived at Delta Facility for his visit. Sitting with a different group of operators, he asked them the same questions he had asked at Alpha Facility. At first, they were hesitant to speak up about the issues they were facing, but John listened to them attentively, reassuring them that he valued their comments.

Their responses were eerily similar. They reported remote controls with worn buttons, missing lifting tools, unclear paperwork, and training that did not cover the skills that the operators actually needed. The same error traps were present, and it was clear that these issues could easily result in another accident.

As John was leaving the site, he overheard the operators talking as they returned to work.

"I can't believe he actually listened to us!" one of them said to the others. "I think our complaints are finally being heard."

Another operator agreed. "It's great to know that someone is finally taking us seriously."

John was pleased to hear such a positive reaction. The operators seemed amazed that a senior director was taking an interest in their feedback, instead of focusing on production targets and following the rules.

••●••

The next day, John put together a report listing the challenges reported by all the operators he had spoken to at Delta Facility:

- The crane remote controls can lose signal, which causes the crane to stop functioning, potentially leading to a dropped load or other serious accident.

- The buttons on the remote controls are labelled with directions such as North, South, East, West but there is no point of reference. This may lead to confusion and mistakes in operating the crane.
- Lifting tools are missing. This causes delays in lifting operations, as operators spend time trying to find what they need, and sometimes they proceed without the right tools.
- Equipment is often not returned from the repair shop on time. This causes delays in lifting operations.
- Operators have not been trained on the new lifting methods described in the official lifting plan. Workers do not know how to safely use the equipment, which increases the risk of an accident.
- Supervisors do not have adequate training to oversee lifting activities. This means that they cannot properly assess the risks, or ensure the operation is being conducted safely.
- Lifting plans are often not available, which causes confusion and increases the risk of an accident.
- Lifting sketches are not always stored with the lifting plans but are often filed in other locations. This causes confusion and delays.

- The documentation does not clearly indicate what to do in the absence of a lifting sketch. This causes confusion and leads to different teams managing lifts differently.
- Information about the load and center of gravity is sometimes missing. This can lead to the load becoming unstable.
- There is no lifting assistance during the night shift.

John presented his findings at the next management meeting.

"A good risk assessment should capture these things," said Brian, the facility manager, sounding irritated.

John paused for a second, frustrated at this response. "Maybe it *should* be captured by the risk assessment," he said, trying to keep the anger out of his voice. "Maybe it should also be captured by our audits, self-verification, or behavioral observations, you name it. But it is *not* captured, and this feedback shows that it has never been captured, despite our efforts.

"The problem here is that you find what you look for. These methods focus on other things. Risk assessments focus on hazards, but the fact that there is no agreed place to store the crane remotes, or the fact that the batteries run out, these are not hazards, but they do increase the risk.

John took a calming breath and looked around the table. "I'm tired of pretending that what we do is enough!" he snapped.

"I'm sick and tired of the tons of paperwork and processes we've generated, just to satisfy legal compliance. Our workers are tired of this approach. They are ready to help, but we are not listening, because we think the regulations don't require us to do this!"

Brian did not respond, and the meeting ended awkwardly.

Chapter 7

CEO update: making the case for change

John sat in his office, staring at the piles of notes and reports spread out on his desk. Over the past few weeks, he had uncovered a multitude of problems, error traps, and challenges that the company's standard processes had failed to capture. Despite the mixed reception from the site management team, he felt a growing confidence. It was time to take his findings to the CEO.

John managed to secure a meeting with the CEO for the following week. As the call began, he immediately noticed the CEO's defensive stance. The CEO's body language was closed off, arms crossed, his tone cold and formal.

"I hope you have some good news," the CEO said, his eyes narrowing as he looked at John through the screen.

"I do," John replied, feeling confident.

"Okay, I'm listening," the CEO answered, his tone guarded.

John took a deep breath and briefly described his efforts over the past few weeks. He explained his conversations with the crane operators at three sites, looking into the challenges of Normal Work, and particularly focusing on the lifting of heavy objects.

He highlighted the significant gap between the issues picked up through investigations, risk assessments, audits, and behavioral observations, and the problems he had uncovered by using open questions and Walk-Through Talk-Throughs with the workers.

The CEO listened attentively, though his hesitation was evident.

"John, in the last two years, we've invested over a million dollars in different safety initiatives and improvements to our risk management system across our businesses. You can understand that I'm frustrated by the recent accidents in your division, and now you're telling me that what we have is still lacking? That's very disappointing on many levels.

"Having said that," he continued, "I'm glad to hear what you've found. It gives me a glimmer of hope that we can improve the safety situation in the company. But ... you know that safety is not the only issue we're facing. Our production is down, and tomorrow I have a call with shareholders. I need to tell them a good story, so our share price doesn't take a hit."

John nodded. "I understand, and I can see how frustrating this must be. But I believe I've discovered something truly helpful that will help not only safety but also quality and efficiency. I would like to ask for your permission to run three pilots at different sites in my division. We can try different approaches to solving the problems and see how we can best make this process work. If it works, we'll increase our ability to address the real safety challenges across the business, and we could explore rolling it out company-wide."

The CEO leaned back, a skeptical look on his face. He remained silent for a few uncomfortable moments.

"Okay, John," the CEO finally said. "You've got four months to come back to me with the results. We need to make a decision. While you are progressing with the pilot in your Division, I'll run another pilot with a different focus in another Division. Four months, John," he said, "and no more. I hope you can replicate your results."

John felt a surge of excitement. "Thank you," he said. "I won't let you down."

Chapter 8

The disappointing pilot

John stood by his office window, his gaze fixed on the bustling activity below. His mind raced with the challenges of initiating his pilot project. So far, only John and a few members of his management team had been aware of his research into Learning From Normal Work. He was excited about broadening the research, but so far most of the organization was still in the dark about his efforts and his vision.

Gathering his team in the conference room, John set up a brainstorming session to decide the best location for the pilot. "We've got sites that perform well and others that struggle, not just with HSE, but operationally and quality-wise too," John explained, setting the stage for the discussion.

The team seemed excited about the pilot project, and they started by discussing the poorly performing sites.

"If this approach can help, then we could give these sites extra support to improve. If we succeed, we'll kill two birds with one stone," suggested Anil, one of the senior team members.

John listened and nodded. Logically, it made sense. Why invest efforts in a well-performing site?

John decided to initiate a pilot program at one of the company's worst-performing sites. If his new methodology was going to be helpful, he should demonstrate its effectiveness precisely where it was most needed.

Later that day, John called Chris, the site manager from one of the struggling sites. He and Chris had known each other for ten years, and he knew he could speak openly. Chris wouldn't talk about their conversation to anyone else.

"This all sounds wonderful," Chris responded when John had outlined his plan. "But we are struggling with the basics here. We don't have enough HSE advisors, there's a high turnover in the workforce, our budget is stretched thin, and last year we were in the red. If you force this pilot on us, sure, we'll implement it, but if I had a choice, I wouldn't prioritize your pilot over the problems we currently face."

John paused, considering Chris's concerns. He was certain the pilot would help solve most of the problems at the site. "Look," he said. "Your site is performing poorly, and you know you need help. Here's the help you need, with the support you need to implement it. Let's do this, and I'll guide you on how to make it work."

Chris reluctantly agreed, and John began planning for the pilot. But despite John's conviction, a nagging doubt remained. Was he setting himself up for failure? If the pilot failed, it could mean the end of his career. He would have to report the failure to the CEO.

••●••

John and his team organized a half-day briefing for supervisors at the site to introduce them to the concept of Learning From Normal Work, and coach them on initiating productive conversations with frontline workers.

"Today, we're not just going to talk about what you're observing," John explained to the gathered supervisors. "We're going to change how we talk about what we see. Instead of using checklists and pointing out rights and wrongs, we're going to ask the workers about problems and frustrations and let them show us the challenges they face."

The supervisors were initially excited. The idea of shifting from a passive observational role to a more engaged conversational approach gave them a feeling of empowerment. This would give them the chance to identify underlying issues and help to address the frustrations their workers were facing.

The early interactions with frontline workers were promising. Supervisors engaged workers in simple conversations, asking about the difficulties they faced. The workers pointed out a number of issues, which the supervisors eagerly brought to the management team's attention, hoping for proactive engagement.

However, the response from the management team was not what they were hoping for.

"These issues they're raising. They're mostly just stupid," one manager dismissed outright.

Another commented, "These aren't the right things to focus on. We know what our problems are."

A third refused to take any action, labelling the efforts as a diversion from real safety issues.

The management team's focus remained rigidly on personal accountability, undermining the initial enthusiasm for the pilot.

"If people don't follow the rules, they will be held accountable," was the prevailing message, which stifled the move toward learning and open communication.

Sensing that the pilot was turning into a failure, John visited the site to discuss progress.

During an official meeting, Chris bluntly stated that the initiative had not only failed to bring positive results but had added an additional burden to an already struggling site. As a result, the management team had decided to discontinue the program.

Deeply troubled by this outcome, John retreated to the canteen for a coffee, and to reflect on what to do next. His career was at stake, and the failure of this pilot could spell the end of his initiatives.

At a table in the corner, John sat with his thoughts, but was interrupted by Martin, a long-time site supervisor, who approached hesitantly.

"Hey, John, can I sit here?" Martin asked, his voice tinged with caution.

"Sure, Martin, have a seat." John gestured to the chair across from him, interested by what Martin might have to say.

Martin settled in and looked around before leaning in a bit. "Look, John, that meeting didn't go great, and they're killing our pilot, right? But there's stuff you need to know."

John pushed his coffee mug to one side, signaling for Martin to continue.

"First off," Martin started, scratching his head. "That training we got? It was only four hours. It's like giving someone a cookbook and telling them they're a chef. We didn't learn anything about what we really needed to do."

John nodded, absorbing the bluntness of Martin's words. "So, the training wasn't enough?"

"Hell no," Martin confirmed. "We were all left kind of hanging, not sure how to actually use what we were shown." Martin shifted in his seat, his voice growing louder. "And don't get me started on the bosses. They didn't get what we were trying to do at all. No clue, and no support. It's like yelling into a hurricane."

John sighed, realizing the lack of leadership support was a bigger problem than he had thought.

But Martin wasn't finished. "John, we're bleeding people here. High turnover, right? We train someone up, and, boom, they're gone. New guys come in, and they know nothing, because they missed the fancy training. When we start asking them weird questions, they clam up. They think it's some sort of trap! It's made them really suspicious of everything we do.

"And the time it all takes, John!" Martin continued. "These chats, they aren't quick. You get into it and there goes the clock. We end up having to stay late to finish our real jobs. It's really annoyed everyone. Got them thinking, 'What's the point?'"

John listened intently, each comment shedding light on the various factors that had led to the pilot's downfall.

"Martin, this is really helpful," John said earnestly. "Thanks for being open with me. I need to think on this and figure out how to tackle these issues."

Martin nodded and stood up to leave. "I just thought you should know what's really going on. We liked the idea, but it's got to be rolled out better than this."

As Martin walked away, John pulled out his notebook, scribbling down his insights.

Each challenge Martin highlighted was a critical lesson for John. He realized that without sufficient training, support from leadership, and clear communication about the purpose and potential of the new approaches, this pilot had been doomed from the start.

Using his notes from their conversation, John sketched out a revised action plan to address the issues. Before leaving the site, he met with Chris and asked him to hold back on communicating the pilot's outcomes. He didn't want the failures on this site to influence his future initiatives on other sites.

The site manager agreed, relieved to shift the burden of the pilot back to John's central team.

John headed back to the office with a renewed sense of purpose, determined to learn from this feedback, and improve the implementation process for future pilots. The challenges were significant, but John was not ready to give up on his vision for a new approach to workplace safety.

••●••

Needing guidance, John called Sarah. "Sarah, my pilot has failed miserably," he began, and then explained the details. "What do you think I should do better?"

Sarah's advice was straightforward. "John, starting with your worst-performing sites is a trap. When introducing something new, you need a feel-good factor, high levels of engagement, and excitement about the initiative. Struggling sites typically suffer from broad challenges that make engagement difficult. Aim to bring them in later, when you have positive feedback from better-performing sites. At the beginning, you need people who are fully focused on implementing your changes, not constantly worrying about wider issues."

"What do you suggest?" John asked.

"Choose a site that has enough resources, time, and capacity to do justice to the pilot, to really learn from it," Sarah replied. "I've said it before, John, don't just tick boxes!"

Chapter 9

Engaging the top-performing site

John reached out to the HSE manager at Epsilon Facility, one of the three top-performing sites. This conversation was very different from his call with Chris.

The HSE manager was enthusiastic. “We’ve been looking for something fresh. We’ve had good results with our audits and observations, and your timing is perfect as we’ve been discussing what our next focus should be.”

The next day, John received an encouraging email from the HSE manager. It confirmed that, after discussion with the Epsilon site manager, they were eager to proceed with the pilot.

However, as John started to plan the details, he realized he lacked answers to some practical questions about the implementation, how much time it would take, how many people would be involved, and other important details.

Realizing the gaps in his preparation, John decided it was time to seek external expertise.

"We need someone with a lot of experience in rolling out proactive learning initiatives," he told his team. "It's time to bring in an expert."

John invited an expert in proactive learning to run a workshop with representatives from various departments at Epsilon Facility, including engineering, manufacturing, and quality control.

A group of twelve people gathered, all eager to understand what John's new initiative entailed.

The expert started by outlining the concept of Learning From Normal Work. He introduced the basic tools and discussed examples and case studies to ensure everyone understood the concepts. The participants, seeking a fresh perspective, were energized by these new ideas.

"As we explore this new approach, it's important to make sure we are fully prepared before starting a full rollout at the site," John explained at the end of the workshop. Everyone agreed, understanding the importance of laying the foundations for a successful pilot.

"We'll start with a one-day workshop for the leadership team to ensure their buy-in and full support," John continued. "Plus, we'll select a group of about ten people from various roles to learn how to identify error traps, and how to use techniques like the Walk-Through Talk-Through and Learning Teams, they will become local experts and champions."

The participants discussed a timeline for the rollout, and the expert John had brought in helped to ensure the key success factors were discussed and agreed on. The site manager was enthusiastic and agreed to cover the costs of the pilot.

A few weeks later, John held the first workshop for the Epsilon Facility leadership team. He called it the 'Mindset Shifter'. Together, they worked through the essential concepts, including adaptations, workarounds, and the importance of understanding work as-done versus work as-imagined.

The leadership workshop was followed by facilitator training, equipping supervisors with the necessary skills to conduct effective Learning Conversations and Walk-Through Talk-Throughs.

Despite the initial engagement and understanding, the first attempts at applying these new concepts didn't yield significant insights.

One supervisor conducted a Walk-Through Talk-Through by talking with another supervisor, rather than a worker. This only captured a polished picture of how tasks were supposed to be done, and the supervisor didn't report any problems.

The expert was involved in ongoing coaching and suggested a change in approach. “Next time, don’t talk to another supervisor. Talk directly to the workers doing the tasks,” he advised.

When the supervisors followed this advice, they discovered many more problems, highlighting the mismatch between the supervisors’ and workers’ perspectives. This was an important lesson for the team, ensuring that the right people would be engaged in future discussions.

Another issue arose when a member of staff attempted to use the Walk-Through Talk-Through method without proper training. The staff member returned frustrated, having only identified typical hazards and complacency, adding nothing to the existing safety assessments.

However, when someone trained in the technique reviewed the same activity, they uncovered issues that the previous person had missed.

This underscored another key point, tools have limited effectiveness without the right mindset and skills.

As John reviewed the team’s initial attempts, he noticed many of the same mistakes he had made himself. He realized the importance of focusing on error traps, rather than emphasizing hazards and behaviors.

Through coaching and feedback sessions with the expert, and through sharing his own experiences, John saw the team's results begin to improve. By the second, and third walk-throughs, both John and the expert were pleased with the quality of the reported learnings.

At a management meeting to discuss the progress, one participant, Safwan, expressed concern about the time-consuming nature of the process. "Each walk-through took between thirty to sixty minutes. That's time away from work," he complained.

John acknowledged the concern but challenged the team to consider the value of the insights gained. "It does take time, but is it a waste? What have we learned, and how has it helped?"

Safwan admitted, "We've learned a lot of useful things. We've made several improvements, and the guys really feel listened to."

"Yes, learning takes time. If we're not learning new things, I agree that we shouldn't do it, but we are. We're identifying issues and error traps, and if we can fix those, we end up saving more time and money and improving employee safety and engagement. Let's discuss with the management team how to account for the time needed as part of the quarterly planning, so people don't feel like they are being dragged away from urgent priorities," John suggested. "Let's think about how we can add this into the site's resourcing strategy."

After the meeting, John asked the team to compile a summary report detailing their findings. He asked for feedback from the participants on how useful they had found the new approach to HSE management.

The team agreed that, while the pilot showed promise for a broader rollout across the company, more thought was needed on how to structure the program to balance useful outcomes with operational efficiency.

••●••

John felt a surge of motivation following the success of his pilot program at Epsilon Facility. Eager to replicate this success elsewhere, he arranged calls with the site manager and HSE manager of the Lambda manufacturing site to present the results from Epsilon.

During the planning meeting, it became clear that the new site was looking for a structured approach, rather than a simple conversation-based learning method.

"We've realized that while our behavioral observation system was helpful, it has reached the end of its usefulness. We're conducting observations, but we're not really learning from them," the site manager explained.

John nodded in agreement. "It's important to move on from observing to actively engaging with the workforce. We need to understand the deeper challenges they face."

They designed the new pilot with three levels of engagement:

1. **Level One:** A direct replacement for behavioral observations. Instead of starting with a checklist and observing safe or unsafe behaviors, they would initiate a conversation about what makes the work difficult. Discussion cards would be used to provide specific prompts. "The idea is that these conversations would be short, no more than ten minutes, so as not to affect production," John noted.

2. **Level Two:** The supervisors saw value in running Walk-Through Talk-Throughs, but recognized the impracticality of pulling a worker away from a production line for an hour. John wrote in the plan, "We plan to use the production line downtime, and send small groups of facilitators to conduct the Walk-Through Talk-Throughs simultaneously." This approach would allow for more thorough conversations about the activities, enabling facilitators to discover error traps without disrupting the workflow.

3. **Level Three:** For high-risk activities involving multiple teams, they planned to use Learning Teams, semi-structured conversations with groups of workers, not just a single operator. "This level will handle complex challenges, where site-wide decision-making is required," John elaborated.

The team agreed to start the training and roll out the redesigned pilot within six weeks.

As they moved forward with these plans, John received a call from Karen, the Avriox Industries Vice President of Operations, who had heard positive feedback about the pilots so far.

"John, I'm pleased with the progress, but I'm wondering how you're measuring success. When we present the data back to the CEO and other stakeholders, we'll need to show them some numbers."

John had anticipated this question. "I've been thinking about that too. Since we're working on sites that haven't had recent accidents, using injury rates as a metric isn't going to work. We're working on potential indicators that would not only show the progress of the work, but also the impact of our efforts," John responded.

He shared his list of indicators with Karen.

- Level 1: percentage of learning facilitators trained.
- Level 1: number of learning conversations (redesigned behavioral observations) per production area and per observer.
- Level 1: number of error traps identified, per observation and per observer.
- Level 2: percentage of production areas with Walk-Through Talk-Throughs and Learning Teams implemented.
- Level 2: number of Walk-Through Talk-Throughs and Learning Teams per production area.
- Level 1 and 2: number of error traps identified in pre-defined categories.
- Level 1 and 2: percentage of error traps eliminated.
- Level 1 and 2: percentage of risk assessments updated.
- Site Risk Profile: percentage of high, medium, and low risks, based on updated risk assessments.

"That sounds promising," Karen responded, clearly reassured by John's proactive approach. "Keep me updated on these indicators and make sure they clearly reflect the success of the new program."

Encouraged by the support from Karen, John felt confident that the pilot would be expanded. He spent the following weeks working closely with team leaders at the new site to ensure they were prepared for the upcoming changes.

They focused on training the facilitators, finalizing the discussion cards, and coordinating the production line downtime to allow for longer conversations.

As the date for the first training session approached, John reviewed the entire plan, ensuring that every detail had been addressed. He understood the importance of this pilot, not just for the safety and efficiency of the site, but for the whole company's approach to workplace safety.

The pilot was not just about changing day-to-day work; it was about transforming the entire culture of the organization.

Chapter 10

Presentation and unexpected approval

Several months later, John gathered his team to review the results of the three pilots. The first pilot had failed, but the subsequent pilots had been more successful.

"We've come a long way, and it's important that we get this presentation right," John emphasized to his team as they brainstormed how best to report their results.

After the team meeting, John crafted a detailed presentation and met with Karen, the VP of Operations, to share his findings. John and Karen decided to make the presentation together.

John's report included:

- Lessons learned.
- Examples of improvements made.
- Interviews with on-site personnel (experts, observers, and workforce).
- A before/after test, a comparison of the insights gained from behavior observations (used before the pilots), and those gained from John's new proactive learning processes.
- The level of risk reduction achieved by using the new processes.
- A comparison of costs between the new processes and the existing safety procedures.

They discovered that the cost of a high-severity accident could vary from tens of thousands to millions of dollars. The total cost of a serious accident might include a range of direct and indirect costs, such as:

- Increased insurance premiums.
- Insurance claims against the company.
- Legal costs.
- Damage to buildings or equipment.

- Lost production due to shutting down the site.
- Employee absence.

The indirect costs would include:

- Fines resulting from prosecution.
- Sick pay.
- Cost of staff recruitment and replacement.
- Production delays.
- Overtime payments.
- Administration costs.
- Training costs.
- Lower morale and productivity.
- Cost of implementing wasted corrective actions that don't eliminate the true causes of the accident.

They also discovered that the costs of accident investigation alone, which could require hundreds of hours of work from investigators, consultants, HR, managers, and legal personnel, plus travel to the site, could reach tens to hundreds of thousands of dollars.

The cost of a Learning Team included:

- Eight hours of training for leaders (assuming ten leaders on the site).
- Sixteen hours of training for facilitators (assuming two facilitators on the site).
- Eight people involved in a discussion for a total of four hours.
- Visits planned in advance to fit around the production schedule.
- Eight hours for preparation, planning, and follow-up for one or two facilitators.
- Cost in time and budget of corrective actions and removing Error Traps.

To their surprise, Karen and John discovered that the combined costs of Learning From Normal Work represented around 0.002 per cent of the cost of an accident!

Thinking back to his own experience, John also realized that an accident could lead to strong emotions among the workforce and managers. Anger, frustration, fear and depression could all affect the productivity of the company, and make workers feel less safe and secure in their jobs.

In contrast, the costs of Learning From Normal Work were far lower. Managers and workers would spend time identifying Error Traps and making changes, but the costs in time and budget, and the emotional impacts, would be minimal.

On the day of the presentation, John and Karen were checking through their notes in the executive conference room when the doors burst open.

The CEO rushed in. "Look," he said, catching his breath. "I know our meeting is supposed to be in half an hour, but there's an urgent issue with a major customer that I need to resolve now." He checked his watch. "I can't stay, but I wanted to make sure I had a chance to speak to you today."

John and Karen stood, surprised, as the CEO continued. "Apologies for the hurry, and for missing your presentation." He pointed at the notes on the table. "I can see you've worked hard on it. I have good news, though."

He smiled. "John, I've heard very good things from those two sites. The site managers had a lot of positive things to say during our management meeting last week, and that's enough for me. So, the next step will be to launch Learning From Normal Work across your Division, John. Only your division, for now. It's the biggest in the company, and it should give us a sufficiently big test for the new approach."

The CEO's words came rapidly. "As I said before, I'm also trying something different in another Division, but I'll support you running this pilot in your division. Send me the plan for the resources you need so we can secure the budget. Based on feedback from the site managers, I'll ask you to work more closely with the quality and manufacturing folks. It seems that your program can help, not only with safety, but also with quality and efficiency.

He took a breath and glanced at his watch again. "If we're making these changes, I'd like to see an integrated approach. I'd like to see initiatives that reduce the risk of injury, while also reducing our quality defects. I want to use these changes to improve our operational Key Performance Indicators as well as our safety record. From what I've heard, your pilot should be able to deliver."

Without giving John and Karen a chance to speak, he turned to the door. "I can't stay, but thanks so much for everything you're doing. I would like to see the plan in a couple of days."

John stood frozen in surprise, watching the door swing shut behind the CEO.

Karen, smiling gently, nudged John back to reality. "Well, that was a success, wasn't it?" she said with a grin.

John felt dazed. The CEO had arrived and departed so quickly, and it took John a moment to process his words. "I suppose so," he said, shaking his head. "But now what? It's a big jump from a pilot to a division-wide roll-out."

Karen picked up their notes for the presentation and handed them to John. "Time to make a plan," she said, still smiling.

Chapter 11

John has no clue how to scale up

John arrived home late, feeling confused and exhausted. Without even giving his presentation, he had received full-scale approval from the CEO, a move he had not expected to happen so fast. His mind had been racing all afternoon, overwhelmed with the plans he needed to make. As he walked through the door, Melissa, his wife, immediately sensed his tiredness.

"Is everything OK?" she asked, noticing his weary expression.

"I'm not sure," John replied. "I need to do some thinking."

"Sit down," Melissa said. "I'll get us both something to drink, and you can tell me all about it."

John settled into the sofa by the window, overlooking the quiet street below. He stared out into the night, watching the cars and the occasional person walking past.

Melissa joined him. "Is it good news?" she asked.

"Yes," he said. "The CEO approved our new approach to safety management. We didn't even give the presentation. He just said yes."

"That's great," Melissa responded, smiling. "Congratulations!"

John nodded. "It is. I'm really pleased. But everything is happening so quickly, and it's going to take a lot of work. He wants me to introduce the program across my division. I honestly don't know where to begin."

Melissa nodded. "It sounds as if you've already started. Not very long ago, you were worried about keeping your job. And now? You have the chance to change things on all your sites. You've already convinced the CEO. What else can get in your way?"

"Well ..." John paused. "Running three pilot programs was manageable. I only had to work with three site managers, and I brought in a consultant to help with the training. Scaling up to fifty-six sites, all over the world? Nearly forty thousand employees? That's going to be a lot harder."

Melissa frowned. "Surely you don't have to oversee everything yourself?"

"That's what I'm trying to figure out." John gave a heavy sigh. "There are some sites that always perform well. I might be able to give them the training and let them run the program without much support. But the others? The struggling sites? They're going to need much more attention, and I'm not sure I can spare the time."

"Can you bring in an external company to give the training?" Melissa wondered.

John nodded. "I've thought about it. We've tried it before, and it is useful to have experts on board. I could send them to deliver the training sessions, and they could help plan the changes for each site." John stared out of the window, lost in his thoughts. "Or maybe I should think about sending people from my sites to external courses, and turn them into experts?"

"That's a good idea," Melissa said. "You'd end up with the people who know the company being the ones introducing the changes. They'd be learning your new methods for safety management, but they wouldn't have to learn how the company works as well. And they'll have more motivation to make the changes work."

"That's true," John said. "The new approach is more likely to succeed if the experts carry on working at the sites. They can make sure the new ways of working are properly integrated with the company's existing systems."

"But ...?" Melissa could see John wasn't happy with this solution.

"But ..." John frowned. "Training my staff to be experts would take time. It would slow down the process of change, and it would cost more than the Division can afford."

"So, you need an affordable solution, that doesn't take too much time and makes sure that the new processes are fully integrated with the way the company already works." Melissa counted the points on her fingers. "And you need to make sure the changes stick. You don't want sites going back to the old ways of working once the training is over."

"And I need to adapt the training to each site, to make sure they have enough support." John shook his head. "I don't know how to make this work."

"These are tough questions, and there's a lot to think about. I don't envy you," Melissa said. "But you're not going to solve this tonight. Come on, let's get some sleep."

As John lay in bed, the questions continued to race through his mind. He knew that approval from the CEO had been a significant victory for his ideas, but he could also see that the real work was just beginning.

John woke up the next morning feeling more drained than rested. He found it hard to keep his eyes open over breakfast, but he knew he had to start making plans for the roll-out.

With no clear answers from the previous night's conversation, he sent an email to Sarah asking for guidance. A few hours later, Sarah responded with congratulations, and sent John several points to consider.

John scribbled them down:

1. Aligning leadership.
2. Integrating with existing processes.
3. Resourcing the roll-out.
4. Ensuring broader awareness.
5. Sustaining the effort.

John stared at the list, wondering what to do next.

Chapter 12

The planning begins

John called a planning workshop for his division management team. He included operational managers, quality representatives, engineering managers, and his HSE team. Each member brought a different perspective to the meeting as they gathered in the conference room, but everyone was keen to work together to scale up the Learning From Normal Work program across the Division.

As the workshop began, John explained that there was no one right way to adopt new practices into an organization. "Every company, and every site, is different. Some take a light-touch approach without much coordination, while others use highly structured project management techniques. We have to find what works best for us."

John looked around the room at his team. "We've got the green light from the CEO, and now it's up to us to figure out the best path forward.

"As we expand this initiative," John continued, "we need to consider how we'll integrate Learning From Normal Work with our existing processes. We need to ensure we have enough people to roll this out, while keeping everyone informed and engaged. It's not just about doing something new to tick a box and look good. We need to take what we already do well and make it better."

There was a brief silence as the team absorbed the scope of the task, and then the disagreements started.

"I think we're overcomplicating this," said one of the quality representatives. "Why not just run the new approach as a separate initiative at first, see how it goes, and then think about integration?"

"That would be a mistake," countered one of the HSE managers. "If we don't integrate this from the start, it'll be seen as another flavor-of-the-month program, and no one will take it seriously. It needs to tie directly into our existing safety and quality processes to have any real, lasting impact."

John listened to the discussion. He knew today's meeting was important. To move forward with planning, the team needed to be open about any doubts they had about the roll-out.

"Let's get everything on the table and talk this through," he said.

Raoul, a seasoned HSE engineer, spoke up. "John, I get why we're trying new things, but a lot of folks are nervous about this. There's a lot of talk about sticking to what's always worked, especially now when safety numbers are shaky."

John nodded, understanding the concern. "I hear you. It's important for us to keep what works, but we should also explore better ways to prevent accidents."

Helen, the operations manager, jumped into the conversation. "Shouldn't we be trying everything to make things safer if we have the chance? If we stick to the old ways, aren't we just waiting for more accidents to happen?"

Raoul shook his head. "It's not just about trying new things for the sake of it. We need to be sure we don't shake things up too much. We don't want to lose the good things we already do."

The conversation jumped between different opinions. Some team members were clearly anxious about moving away from methods that seemed to work, while others felt frustrated, wanting to push for greater innovation.

John saw the need to guide the team toward a middle ground. "What if we blend the old and the new?" he suggested. "We could use these new ideas to fill in the gaps where our current methods aren't good enough. This way, we're building on what we already do well."

A data analyst added, "We should watch the numbers closely as we try this out. If something isn't working, we can always pull back or try a different tack."

John was encouraged by this turn in the conversation. "That's a great approach. Let's keep a close eye on how things go, and make sure we're not making anything worse. But don't forget, if we don't integrate the new approach, we risk losing our momentum and any connection to our core operations. These concepts and tools need to be woven into the fabric of what we do every day."

The discussion turned to resourcing. "We can't expect to roll this out without substantial support. We need dedicated teams, not just volunteers who already have full-time responsibilities," argued an operational manager.

John agreed. "We'll need to identify and train regional and site-level champions. They'll be our front line in implementing these changes, and they'll need to fully understand the process of Learning From Normal Work."

The team discussed the number of champions they would need to train. They estimated that the roll-out would involve fifty-seven site managers, around forty corporate leaders globally, and approximately two hundred and seventy site-based leaders across quality, operations, and engineering.

A team member raised a concern. "If we're imposing this on all the sites in the Division, isn't there a risk it might fail if the leaders we've identified don't buy in? How do we make sure they're prepared and supportive?"

John responded thoughtfully. "That's a great point. We need to make sure they are supported and ready to roll out the changes. Let's think about scalable ways to do this."

The team agreed on a three-tiered approach for engaging leaders.

1. **Email communications:** a series of emails created with the help of the communications team to introduce what's coming, explain why it's coming and highlight examples from the pilot projects.

2. **E-learning modules:** online training to provide an introduction to proactive learning and practical tools.

3. **Leadership workshops:** in-depth sessions throughout the year to help leaders move beyond error and non-compliance and see behaviors as adaptations to error traps as well as develop foundational questioning skills so they can lead by example.

"Additionally," John added, "we'll ensure that our quarterly global leadership meetings include a section for discussing this roll-out. We'll let the leadership team know what's happening, and we'll give them examples of the positive changes we're making."

The next challenge was to decide who would introduce the changes to the workers.

"We have about two hundred HSE specialists across our sites. Not all sites have them, and those that don't will need additional support," John noted. "We also need to be aware of cultural differences across the Division that might need special attention."

The team decided to invite the site HSE specialists to collaborate on designing workshops tailored to their unique geographical and operational contexts.

"We also need champions who are not only skilled but can also advocate for this initiative and manage any pushback," John stated.

The team came up with a plan to train a select group of HSE specialists to become champions for the roll-out. They would be given the skills to lead the learning activities and support other sites as necessary.

John then detailed the three levels of implementation he had developed for the pilot program at the Epsilon Facility.

1. **Enhanced leadership conversations:** Supervisors would be trained to use proactive learning questions as part of their regular safety observations.
2. **Scheduled Walk-Through Talk-Throughs:** In-depth reviews would allow supervisors to fully understand the work practices in each part of their site. Reviews would be scheduled in advance to avoid disrupting production.
3. **Learning Teams for major challenges:** More complex issues would be addressed by involving more workers in the reviews and developing an understanding of the entire production process.

"But what will happen to all the findings from the reviews?" asked a team member.

Helen spoke up. "For the insights we gather, we'll need a clear process for making changes. This means checking with the relevant departments to find out whether the issues we find truly affect operations, and whether any proposed solutions are workable."

Stuart from procurement added, "Once we approve these solutions, we must ensure they're resourced properly. This might mean adjusting our procurement strategies, or reallocating budgets to address the issues that come up."

Yasmin, Head of Quality Assurance, suggested a solution for sharing information. "To capitalize on what we learn, I propose we set up a shared digital platform. This would allow sites to upload their findings and solutions, making it easier for all sites to access and learn from each other's experiences."

John nodded in agreement. "Great idea, Yasmin. We should also organize quarterly cross-site learning sessions, where teams can share their successes in person."

John then steered the conversation toward how they would monitor and evaluate the impact of the new safety practices.

Morten, the data analyst, spoke up. "We should track the roll-out using progress indicators. So, for example, the number of walk-throughs conducted each week, or each month would give us helpful feedback on how the new processes are being introduced."

Nala, the senior HSE specialist, highlighted the need for deeper impact measurements. "Beyond the basic indicators, we need to assess whether these new practices actually reduce risks."

"We should also include operational efficiency and quality measurements," added Alec, the engineering lead. "If we're doing this right, we should see not just safer environments, but also more streamlined operations."

John encouraged the team to consider how they could maintain a continuous feedback loop. "How do we make sure these indicators are reported, and how do we make sure site teams act on any feedback from experts and managers?"

Jenny from corporate communications made a suggestion. "We could set up an internal monthly newsletter that reports on the progress of our safety initiatives. We'll make sure we celebrate successes and encourage open discussions of the challenges."

"Quarterly reviews with all stakeholders would be useful, too," said Mike, a regional manager. "We should present our metrics but also gather informal feedback from the workers and champions. That way, everyone can see what we're doing and exactly how it affects safety in the company."

The workshop continued with discussions on the training and certification needed for the champions. They explored several options, and decided to enroll their champions, plus members of the site and Division leadership teams, on the Learning From Normal Work Competency Pathway, an in-depth course covering a range of topics and tools to enable the champions to lead the changes at their sites.

The Competency Pathway covered a number of topics, such as techniques for asking effective questions, how design leads to errors and workarounds, how to predict human error, the psychology of procedural non-compliance, the limitations of using safety metrics, and many more.

The training was structured as a blended course:

- **Virtual classes:** these sessions would cover the ideas behind Learning From Normal Work, alongside examples, case studies, and individual learning assignments.
- **On-site activities:** these would enable the practical application of learned concepts through templates, checklists, and direct engagement with site stakeholders.
- **Assessment and certification:** participants would have to pass an online knowledge test and demonstrate practical application skills across all topics to receive a certification. They could then effectively lead the changes to HSE management on their respective sites.

This training would ensure that champions were not only trained in theoretical knowledge, but also skilled in practical applications, ready to lead and support Learning From Normal Work across the organization.

The team agreed to implement the training and certification on a select number of diverse sites. This would allow them to test and refine the approach, then expand the program across all sites, supported by comprehensive up-to-date resources for training and communication. John also enrolled on the Competency Pathway in order to effectively lead the roll-out.

As the workshop wrapped up, the room buzzed with excitement. The team had mapped out a solid plan while still recognizing the challenges ahead. John thanked everyone for their contributions.

"This is the start of something big," he said. "We have a chance to not only improve safety, but also to improve our operations across the company."

••●••

Later, John reflected on the day's discussions. His notes from the meeting showed a lively conversation.

- Should we integrate the roll-out with our existing systems, or wait to see whether the new approach succeeds?
- Is this the right time to introduce a new approach, so close to the two serious accidents? Should we take the opportunity to make changes while safety is fresh in

our minds? Or are we jumping into something new for the sake of it?

- We need to monitor the results to see how the program actually affects health and safety.
- We need to make sure we have the resources to run the roll-out, train dedicated teams, and include corporate leaders, site managers, and site-based leaders across departments.
- Support leadership buy-in with a digital reporting platform, a monthly newsletter, and quarterly review meetings.
- Use the Learning From Normal Work Competency Pathway for certification.

John frowned as he remembered the disagreements. Some of the conversations had been tough but necessary, and they had sparked the critical thinking needed to shape the plan for change.

With a sense of relief, John appointed a project manager, Nadiya, to implement his team's plans, and began to work his way through the Competency Pathway training. If he was going to bring the company along with him, he needed to lead from the front.

Chapter 13

Where safety science hits the wall of practical application: the problem of investigations

John leaned back in his office chair, his gaze drifting past the computer screen to the rain streaking down the windows. The rhythm of the rain mirrored the hum of thoughts in his mind.

Through his research on Learning From Normal Work, John had uncovered insights he had never reached using traditional accident investigation techniques. The realizations had been profound, yet unsettling.

"Why is it that both success and failure stem from the same factors, but traditional investigation methods only look at failure?" John mused.

The Walk-Through Talk-Throughs and Learning Teams had opened his eyes to the daily realities that his workers faced. He had been surprised by the subtle yet significant error traps that rarely led to accidents, thanks only to the adaptations and workarounds used by the workers. He realized that these everyday challenges were not picked up through investigations after an accident, but that the adaptations were often demonized through judgmental labeling.

Despite the progress he had made using proactive learning methods, John's dissatisfaction with the accident investigation process gnawed at him. Typical investigations after an accident, often ending with conclusions like 'human error' or 'complacency', felt superficial and unhelpful.

"What's the context driving these decisions and behaviors?" John pondered, searching for a way to capture the real situations that pushed employees toward unsafe adaptations.

Driven by this need, John had begun an extensive review of industry literature on system-based investigation tools. His desk was littered with papers and books, each promising insights, but none offered the practical tools he desperately needed.

Advanced system-based tools held some potential. They were used to trace causal relationships within complex systemic failures, and they were academically respected for their intellectual sophistication. But they also used inaccessible academic jargon, and John realized this would alienate his operational staff rather than enlighten them.

"It's as if there's an invisible barrier between what can be analyzed, and what's practical," John thought, his frustration mounting. He knew that the gap between academic theory and operational practice was not new, but this division seemed particularly wide.

The more he read, the more disillusioned he became with the prospects of translating these complex theories into tools his team could use.

"How do we make these sophisticated ideas accessible to our supervisors and safety advisors?" John questioned, feeling overwhelmed by the challenge.

The problem was clear, but the solution was still out of reach. He needed a bridge between the theoretical and the practical.

He called Sarah, his mentor, and was delighted to hear she was in the area visiting one of her sites. She invited him to join her for the visit.

John parked his car in the site parking lot. He was looking forward to catching up and gaining more insights from Sarah's experience. She greeted him with a warm smile at the main reception.

"John! How's everything going?"

John chuckled as they waited to pick up their security badges. "It's been a whirlwind! I'm very happy that you invited me to join you."

"Me too," Sarah replied. "I'm looking forward to hearing about your progress."

"I hope you have plenty of time!" John laughed.

The security guard placed their badges on the desk. John thanked her and clipped his to his jacket.

"Follow me," Sarah said, leading the way along the corridor. "And tell me more. What's been on your mind?"

"Well," John began, unsure where to start. "I look at the reports from our investigations, and it's all 'human error' this and 'complacency' that, but that doesn't fit with what we're finding when we talk to workers. Honestly?" He shrugged. "The results of my investigations feel like blame, blame, blame. As if we're not able to move past this invisible barrier of finger-pointing."

Sarah pushed open the door to a break room and gestured for John to follow her.

"Take a seat", she said. "Can I get you a coffee?"

"Thank you." John sat down at one of the tables. "You know, the more I look at the results from our pilot programs, the more I realize how much I missed before, when I was using the accident investigations I was trained to do."

"That sounds familiar," Sarah replied as she filled two mugs with coffee. "What's bothering you about your investigations so far?"

John sighed. "I feel as if I haven't gotten to the heart of how our organization contributes to these incidents."

"And that's what you're aiming to improve?" asked Sarah rhetorically, handed him a mug of coffee and took a seat across the table. "I went through something similar a few years back. It feels as if you're only seeing the surface ripples and not noticing the currents below."

"Exactly!" John responded. "I know I'm missing the important information."

Sarah took a sip of her coffee. "What are you thinking of doing about it?"

John thought for a moment, eager to explain his ideas. "I want to modernize our accident investigations," he explained. "I want them to show how our system works, rather than just noticing one or two, maybe three root causes."

Sarah smiled. "Let me tell you about an incident we had here last year."

John nodded, and Sarah continued.

"It began with the alarm sounding. A Level 2 alert, everyone evacuated. Thankfully, no one was injured, but we'd had a spill of butanone which had spread across the floor and begun to evaporate, so we needed emergency ventilation to clear the flammable vapors."

"I'm glad everyone was OK," John said.

"Me too," Sarah agreed. "And we certainly didn't want it to happen again. Once the situation was under control, we started an investigation. The old process would have involved interviewing the technician who had been operating the machinery and finding out whether procedures had been followed."

"But that's not what you did...," John curiously commented.

"Right," said Sarah. "Instead of searching for the causes, we started exploring the context."

"Context? What does it mean? Were you looking for the error traps that led to the spill?"

"Yes, error traps and constraints in place create context within which the operators work. These constraints are created by how the organization is structured and resourced. I'll tell you more in a second."

"You were exploring the wider organization behind the spill," John said. "Looking to see how the systems in place on the site affected the risk."

Sarah gave him an encouraging nod, and he continued. "I've been reading about systems thinking, but it feels really complicated."

Sarah smiled, recalling her own journey. "Oh, I remember that phase."

He shook his head. "The tools I've found to help with this are ... not for me. I can't see how to apply them. They feel so academic and completely detached from what we're actually doing. I can't figure out how to turn the theories into tools that my team can actually use."

"You're looking into system-based tools?" Sarah shook her head. "I'm sorry to say that they're nearly impossible to use without employing a team of specialists."

"Thank you!" John exclaimed, relieved. "It's good to hear you say that. I thought that maybe it's just me. I'm looking for something far more practical. Something my team can actually use without needing a PhD. How did you manage to bridge that gap?"

"Not easily!" Sarah leaned back in her chair. "Initially, we tried to find the right investigation training, something that would take all these things into account, but we couldn't find anything suitable. We looked at Root Cause Analysis training, but it was either too simplistic or far too complicated. We found a few courses that included human factors, but they only did so in passing, not as a key part of the training."

John frowned. "That's frustrating. I thought about adapting some of the approaches to suit my team. Is that something you tried?"

"We did try to adapt some of the courses, and we even tried combining them, but this made our investigations so complicated that we found they were useless in practice. When we presented the results to our colleagues, they were mostly confused and frustrated.

"The thing is," Sarah explained, "the findings were usually accurate, but we needed a simpler way to both reach those conclusions and explain the process we were using."

"So, what did you do instead?"

"Ah," Sarah said with a secretive smile. She took a slow sip of her coffee, her eyes lighting up with enthusiasm. "The first step was to get rid of the term 'root cause'."

John sat back in his chair, shocked at Sarah's words. He couldn't believe what he was hearing. "But that's the basis of all our investigations! Finding the root cause!"

"And there's the problem," she explained, gently. "Problems in large companies just don't have neat-and-tidy root causes."

"But all my safety training ..." John began. "It's all been about finding one, or a handful, but causes nevertheless, and doing something to stop them happening again!"

"And how is that working out for you, John?" Sarah asked.

He shrugged, thinking about his investigations so far.

"Badly," he admitted.

"You said it yourself. Your investigations feel like excuses." Sarah leaned her elbows on the table. "What if they didn't? What if you could figure out what was really going on, and change things to make the whole organization safer?"

"So, I get that you decided not to think about root causes." He shook his head. "I still can't quite believe that it's a very big decision! How did you move on from there?"

"Well ... we had to rethink our approach entirely. We decided to focus on the *context* surrounding the situation rather than the *causes*. Things like the way different teams interact with each other, and how company-wide issues contribute to individual accidents and failures.

"Take our chemical spill as an example," she continued. "We were looking for the human story behind what happened.

When I spoke to the technician involved, he was visibly shaken. He'd followed the same steps he used every day. He kept saying, 'I swear, I did everything by the book.' He wasn't being defensive; he really meant it. So instead of handing him a witness form, I sat with him for a structured conversation to unpack error traps and context."

John raised an eyebrow. "It sounds as if you made a significant change to your investigation procedure. How did you manage it?"

"Well, for one thing, we made the effort to move away from looking for rule violations and using judgmental language."

"Another big change," said John, nodding with some disbelief. Determining compliance was always one of his first steps in the investigations.

"It was, but it was important. We realized we needed to understand the *purpose* behind adaptations and workarounds, as well as understand the wider context."

"So, figuring out the advantages of actions that, in hindsight, we would consider unsafe? Is that what you mean?"

"Right," Sarah nodded. "And we needed to explore how error traps are created within our organization, as well as look at how our different teams and departments depend on each other in creating these error traps."

John was intrigued. "How did understanding those dependencies help?"

"It helped a lot. By mapping out relationships and dependencies across our teams, we could see how risks were co-created. One worker's decisions on the front line didn't happen in isolation. Different teams and departments were making decisions that affected each other."

John nodded. "That makes sense. If one team makes a decision, it could ripple through the organization and create risks elsewhere, right? Last year, on one of my sites, the maintenance department wanted to cut their spending, so they reduced the maintenance team by twenty per cent."

"And what happened next?" Sarah asked with a smile.

John shrugged. "To begin with, nothing. They thought they'd made a great decision!"

"Of course they did!" Sarah laughed.

"But after a couple of months, there was a steady increase in the backlog of maintenance jobs. Equipment wasn't being repaired, so production was being delayed. Supervisors ended up putting pressure on the workers, so the workers started using manual overrides on the machines to speed up the work."

"Let me guess," Sarah said. "The safety teams reported more unsafe working, and the workers were blamed."

"Spot on. Morale dropped. The workers felt as if they had been punished for doing the right thing, when all they were trying to do was speed up production."

"That's a great example," Sarah nodded. "It's a ripple effect. The decisions made in one department can change the flow of resources and information across the business, in ways that aren't always immediately obvious. They might be simple decisions about allocating people, tools, time, resources but they could all affect other departments."

"So," John replied, reflecting on Sarah's words. "You're saying that if you understand in advance how these isolated decisions change things for other people, you can proactively reduce the risk. I like this way of thinking!"

"Yes, but this isn't just about preventing another similar accident. We realized that finding hazards and error traps could only provide part of the picture. We hadn't been looking at things like the flow of information, or resource allocation decisions. We had to rethink our approach to risk management.

"With the chemical spill, on paper, the procedures were solid. The lockout-tagout checklist was signed off. The line purge had been logged. But what we learned during the conversation was that over the previous month, batch schedules had shifted due to supply chain disruptions. Technicians were expected to swap chemical lines several times during a shift, something that used to happen once a day, maximum."

"That doesn't sound good," John commented.

"It wasn't. And when we mapped the error traps, we saw that the equipment wasn't designed for frequent product changes. There were no visual indicators showing which chemical was in the pipe at a given time. The technician had purged the line, but it was the wrong line, and he had no simple way of checking."

John took a sip of his coffee and considered this suggestion. "It sounds as if you had to really challenge simple cause-and-effect thinking."

"Absolutely," Sarah said with a nod. "The idea of a root cause with a straightforward effect isn't helpful when you start to think about how things work in a complex organization. So many relationships go unnoticed when you take a narrow, root-cause view. It's important to remember that accepting and addressing something we *think* is a root cause could actually limit our ability to prevent similar accidents."

"So, a more general view of the organization can provide better opportunities to prevent future accidents?"

"Right. After the chemical spill, we put together a systems map to see how other parts of the organization had influenced what happened. It showed us that Planning had changed batch sequences without informing operators, instead relying on digital dashboards that weren't synced with the floor's manual labeling system. Procurement had sourced similar-looking chemical drums because of supplier issues. And Maintenance had removed color-coded piping tags for repainting and hadn't put them back. No one had done anything malicious, but together, these decisions created a perfect storm."

"The ripple effect," John said.

Sarah nodded. "Precisely. Developing that general view allowed us to develop more effective strategies, rather than finding someone to blame. We got better at reducing risks, and we managed to improve overall safety."

"That sounds like a major change," John said, shaking his head.

"It was," Sarah continued. "But the hardest part was changing the mindset of our investigators and managers, like you, John!" Sarah said half-joking but with a smile. "We had to move away from *blame* and toward a genuine *curiosity* about how our departments really function together."

"And how did you make that change?" John asked, sitting forward and leaning his elbows on the table. He could see that this approach might help him to understand the complex influences that made accidents happen.

"We developed a series of workshops. They were designed to be hands-on, using real-life, practical examples to help our teams understand system-based thinking, but without presenting it as such."

John gave an enthusiastic nod. "That's exactly what I'd like to do!"

"And we also looked out for variability. Differences in how people do their jobs at different times, depending on the exact conditions."

John nodded again. "Like lifting a load with a crane? That can involve completely different challenges on different days, even if everything seems the same at first glance."

"That's part of it," Sarah agreed. "But we also looked for variability in resources or planning."

"So how did it go?"

Sarah finished her coffee and put the mug down on the table. "It was difficult," she admitted. "We were trying to transform our entire investigation system. One of the first things we looked at was the way leaders responded to incidents. In the past, there had been a lot of pressure to complete our investigations quickly, sometimes within just a few days. The leadership team demanded quick answers, and that pressure didn't help us to learn from what had happened."

John frowned. "That's exactly how I felt during our investigations. How did you change things?"

"We worked with our managers to help them understand what a useful response would look like. We showed them how to provide time and space for people to learn from incidents, instead of rushing to find easy answers. And we recognized that physical injury isn't the only harm people experience, there's also psychological harm."

John remembered the paramedic who had told him about the impact of Harry's accident on Dan, the crane operator. "You mean like anxiety or trauma after an incident?"

"Exactly," Sarah said. "It's not just the injured person who might be affected. Witnesses can be traumatized by what they saw, and even managers can experience anxiety or post-traumatic stress disorder."

"But what can you do about that, other than giving people time off to recover?"

"Well, we introduced psychological first aid to complement physical first aid. We also replaced traditional witness statements with impact statements."

John raised an eyebrow. “Impact statements? How are they different?”

“Ours include questions about each person’s subjective understanding of the situation. Like a witness statement, but more personal. People involved in the incident tell us about the impact it had on them and their colleagues and give us their thoughts on potential improvements. It’s a great way to learn from what happened while the experience is still fresh in people’s minds.”

“That’s smart,” John said, impressed. “It must have taken a lot of time to develop the new system.”

“It did. We realized that just asking people to report the events they’d witnessed wasn’t enough. The prompts we used to start with didn’t give us the insights we needed.”

“So? What did you do?” John asked, ready to hear her answer.

“We’ve been working with an external expert to develop a two-tier training program. One for our master investigators, five days of hands-on practice, and another, two days long, for a broader group of supervisors and managers. This training gives them key skills, like how to usefully describe context, to use neutral language, and to employ a range of inquiry techniques.”

John was silent for a moment, staring into his coffee mug, absorbing everything Sarah had shared.

"I'm impressed," he said, after a moment. "It sounds as if you've really transformed the way your organization approaches investigations. You're not just finding out what went wrong on one occasion. You're developing a practical understanding of the system, and you're using that to make meaningful changes."

Sarah nodded. "That's exactly right. It has taken a lot of work, but the results have been worth it. We've created a culture that prioritizes *learning* from incidents, instead of finding someone to *blame*, and that's made us stronger as an organization."

She smiled. "Shall we take a walk? I'll show you some of the results of our changes."

"Of course," John replied. "I'd like to see some of your insights in action!"

Chapter 14

Changing the company culture

Sarah picked up a clipboard and led the way out of the break room and along the corridor.

"One of the significant changes we made," she explained as they walked, "was in the language we use around investigations. We stopped talking about 'investigations' altogether, and started focusing on learning, and partnering for learning."

John was puzzled. "No more 'investigations'? Why the shift in language?"

"It's about changing the mindset," Sarah said as they headed outside. She led the way into another building and up a set of stairs, onto a walkway over a busy workshop. She raised her voice to be heard over the noise. "'Investigators' were seen as people who came in after something had gone wrong, looking for someone to blame. This mindset filtered the information they received, and often led them to focus on engineering problems or on blaming individuals."

She stopped and held out her hands towards the work going on in the workshop. "What you look for is what you find, John!"

John nodded, looking down at the workers. "If I'm looking for someone to blame, I'm more likely to find someone to take the blame."

Sarah nodded. "With our old model, the investigation into the chemical spill would have blamed the technician for selecting the wrong line, even though he had followed the correct procedure. But instead, we asked how the system had allowed this to happen. Instead of writing up new rules or sending out a 'be more careful' memo, we brought the team together for a restorative dialogue session."

"Everyone in the team?" asked John.

"Everyone. The technician described what he saw, what he thought, and why it made sense in the moment. Maintenance admitted they hadn't prioritized re-tagging because the job wasn't classified as critical. Planning agreed their new batch schedule process had bypassed the communication chain."

"So you made real progress by listening to the people involved."

Sarah smiled. "That's right. We realized that if we wanted to truly *learn* from accidents, we needed to move toward partnerships and collaboration, and away from immediate blame. Not only that, but I've seen research showing that the *words* we use actually determine the *logic* that we use."

John gave her a puzzled look. "What do you mean by that?"

"Well, for example, if I say, 'Matt violated the safety rules,' I'm more likely to blame Matt, compared to saying 'Matt adapted to the situation'."

John nodded slowly. "So instead of treating an incident like a crime scene, you're approaching it more like a collaborative effort?"

"Exactly," Sarah confirmed. "In the past, we did think of accidents as crime scenes, and investigators were looking for 'criminals' who might be hiding information. But an emphasis on *learning* instead of *blame* helps us to see everyone involved as partners with different roles and perspectives."

John glanced down at the busy workshop, and the people working side by side. "That makes a lot of sense."

"When we brought everyone together to discuss the chemical spill, they all had a chance to understand the system they were part of and to work to improve it. Together, they proposed adding real-time visual indicators on mixing lines, creating a single shared planning dashboard visible from both the office and the floor, and reviewing job classifications for maintenance based on risk, not aesthetics."

John was impressed. "So they all worked together to solve the problems?"

"That's what's so powerful about this way of looking at investigations," Sarah explained. "As opposed to the small team of investigators coming up with isolated improvement ideas that are disconnected from resource constraints, information flows and team dependencies."

She looked down into the workshop. "We also make sure we keep up ongoing conversations with operators and other personnel. And it's not just to gather information, we keep them in the loop, update them on our findings, and ask for their input throughout the process."

John's eyes lit up. "So it's a continuous dialogue, not just a one-time interview after the accident?"

"Yes," Sarah said with a smile. "We involve everyone in the shared process of finding out what happened. We recognize that no one person has the full picture, and maybe finding the full picture isn't even possible! By making connections, and encouraging people to work together, we can gather better information and help build trust and collaboration. We've not only improved the investigation process, but we've also helped people to work together again afterwards."

John was impressed. "You're creating a culture of safety and learning at every level."

"That's the goal. After an accident, we now ask about what the person involved was trying to achieve. We ask about the error traps they faced, how they adapted to those challenges, and what made sense to them at the time. Even if someone took an unsafe action, we discuss the advantages they saw in that moment. We've learned that most people aren't trying to be dangerous!"

"I suppose that's true," John agreed, thinking about the accidents on his sites. "When something goes wrong, it's probably because someone is just trying to get the job done, in spite of the challenges they're facing."

"That's important to understand, but don't forget, that's not the only information we look for. We also look at the broader organizational processes that contribute to the error traps."

John raised another question, gesturing toward the busy workshop. "It all sounds great, but how do you make sure this doesn't become just another time-wasting box-ticking exercise?"

"That's a fair question," Sarah replied. "The investigation framework we've developed provides structure for the investigator and helps uncover different aspects of the accident." She counted the list on her fingers. "We look at expected procedures, available resources, task design, stress, supervision, information flow, and so on. For more complex cases, we can use decision analysis to understand the appeal of decisions that might seem unreasonable in hindsight. This often reveals that those decisions made sense, given the context at the time."

John folded his arms, watching the workers as they concentrated on their tasks. "So how do you capture all this information in a way that's useful for everyone?"

"Ah," Sarah replied. "We create influence maps. For example, if an operator selects the wrong tool," she said, pointing out people in the workshop below to emphasize her words, "we might trace back and find that an engineer provided incorrect information, a supervisor assigned an unfamiliar task, and a manager agreed to an unrealistic deadline. Just like the chain of influences we uncovered after the chemical spill."

John nodded. "And what about correcting the problems? How does that work?"

"It's important to keep taking a wider view of the system as a whole," Sarah explained. "We try not to fix individual issues in isolation. It's tempting to think you've solved a problem if you control a single hazard, but that makes it harder to see the larger issues, and you're back to identifying a single root cause. Traditional hazard controls and corrective actions can still be helpful, but we recognize that they only cover part of the picture."

"And how do you make sure your big-picture corrective actions are implemented properly?"

Sarah smiled. "This might be the most interesting part. We make sure any changes and corrections are developed and reviewed by both technical experts *and* the end-users. Any corrective action has to be signed off by the experts, *and* by the person who will be expected to implement it."

"So the workers have to agree with your changes to their tasks?" asked John, surprised.

"Of course. If they don't find our changes useful, they'll decide not to use them, and we'll end up with more hidden workarounds and error traps."

John nodded his approval. "That makes sense."

Sarah shrugged, pointing again at the workshop. "It's essential. Our main aim is to make it *easy* to do the right thing, and *hard* to do the wrong thing."

She continued. "Using our old process after the chemical spill, we would have recommended disciplinary action for the technician, but that wouldn't have been a fair outcome. Instead, we recommended redesigning the system to make it easy for everyone involved to do the right thing."

John nodded. "That's a really big change."

"It is. The technician who spilled butanone is on our cross-functional improvement team. Maintenance now uses QR-coded tags. Planning has a daily sync with operations. And we've introduced peer-led incident reviews that include people from every affected department."

"Everyone working together, trusting each other rather than hiding what went wrong? That's a powerful outcome," John said. "And it makes a lot of sense when you're on the front line, like this, and you can see the way things work in reality."

He thought about the CEO, and his expectations for the roll-out in John's Division.

"I need to ask," he continued. "What do the leadership teams think about these changes?"

"Well, to reflect the change in culture, we had to change how we write reports. We need to communicate what we've discovered effectively to the leadership teams, without simplifying our findings. So instead of listing failures, we now use storytelling."

John gave her a puzzled look. "Storytelling?"

Sarah nodded. "We create narratives from the employee's perspective, what obstacles they faced, how they worked to overcome them, and so on. This provides context, and it helps leaders empathize with those involved. You remember when

I said I sat down with the technician involved in the chemical spill, and asked him how the day unfolded? That helped me to understand the narrative from his point of view. Our final report described the accident not as a timeline of failures, but as a story of interactions. We explained that the event was not caused by one wrong decision. Instead, we demonstrated that a series of individually reasonable choices led to the final outcome. These reasonable choices, when combined, led to a dangerous situation that no one person could predict."

John was impressed. "Wow. That's an amazing transformation. It must have taken a long time to bring everyone round to a new way of working."

"It took us about two-and-a-half years to fully make these changes. We had to train people and change our processes. But today, investigations are a much more positive experience for everyone involved, even when the accident is extremely serious.

"We emphasize making sense of the wider situation, instead of focusing on mistakes. We amplify the voices of everyone involved rather than silencing them, and we don't try to fix people. We've switched to understanding how the system shaped their choices, and that's what produces a positive outcome."

John took a deep breath. "This is all really impressive. But I'm still not sure how you handle accountability with this new approach."

Sarah stepped away from the railing, her expression growing serious. "That's a challenge," she admitted. "We've moved toward a restorative justice approach. We recognized that if we humiliate or shame people after an event, workers are more likely to hide information or even sabotage our efforts to find out what went wrong."

John nodded. "They're used to the old system, so they'll be expecting punishment."

"Exactly. So instead, we focus on making people feel appreciated and respected. We understand that they might feel bad about what happened, but we don't want to isolate them. We want them to be *more* engaged after an incident, not less."

John looked out across the workshop. "And how do you achieve that?"

"We use restorative dialogue," Sarah explained, continuing along the walkway and beckoning for John to follow. "After the investigation is complete and things have settled, we bring everyone together. Each person gets a chance to express how they were affected. They get to tell us what they thought when they realized what had happened, and to share anything they believe could make things better."

"That sounds great, but how do you make sure the workers engage?"

"You asked about leaders earlier," Sarah replied. "And the role of leaders is critical for this stage. If they are aggressive or blame-focused, you're right, the whole process falls apart. But when they are properly trained and prepared, we see transformative effects. Workers become stronger advocates for safety when they feel involved."

John walked in silence for a moment, absorbing everything. "It sounds as if you've separated accountability from the investigation process."

"We have," Sarah confirmed, pulling open the door to another corridor. "We don't shy away from using punishments if needed, but our main aim is to make sure there is clarity, around the rules, and around any corrections we have introduced."

"It's more important for everyone to understand how to move forward safely," said John, reaching out to hold the door for Sarah.

"That's exactly right," she confirmed. "And our investigators are on site to learn and understand, *not* to punish. Accountability is handled separately. This approach helps us maintain a positive, engaged workforce even after something goes wrong."

"There is so much more I could be doing." John shook his head as he let the door close behind them. "And I imagine keeping all these new skills sharp is a challenge, especially if you don't have many accidents to investigate."

"Right again, John," Sarah said, heading for another set of doors. "It's a big risk for any organization. It helps that we've integrated Learning From Normal Work into our investigations."

"How does that help?" asked John. "It's something I've been looking into."

"Well, we use the same tools, worker interactions, and approaches for pre-incident learning and post-accident investigations. Our investigators are using their skills all the time, and that helps us to keep improving our understanding of how to work safely."

"That makes sense," John said, feeling both inspired and overwhelmed by their conversation so far. Sarah had given him a lot to think about, and he was excited about applying it at Avriox Industries. As they headed through the doors and down another set of stairs, he was already thinking of new questions to ask.

Chapter 15

Re-imagining risk assessment

"You've made so many changes," John said. "What else have you transformed?"

Sarah paused at the bottom of the stairs, unclipped a sheet of paper from her clipboard and handed it to John. "Here's our updated risk assessment form. What do you notice?"

John looked at the form in his hand. "You're not just looking at hazards," he said after a moment. "I see a column for error traps."

Job Steps	Potential Hazards	Hazard Controls	Error Traps	Error Trap Optimizations

"Exactly," Sarah confirmed. "This is something we noticed early on. We were using standard risk assessments, but like you, we were still seeing accidents. In some cases, the accidents were repeating themselves. Our traditional risk assessments focused on identifying hazards. As for error traps and adaptations, we were clueless. We didn't have a way to tell the difference between work as-imagined and work as-done."

John asked. "Can you give me an example?"

Sarah smiled and pushed open the door at the bottom of the stairs. John followed Sarah onto the noisy factory floor.

"This is a great example," she began, raising her voice and pointing. "An older piece of equipment where no one had thought about the error traps." She led John through the busy space to a control panel against a wall. "What do you see?" she asked.

John studied the control panel in front of him. A set of pipes ran vertically in front of the control unit, blocking access to the buttons behind. "Those pipes make using the panel much harder," he commented.

Sarah smiled. "You're right about that. The worker needs to press buttons on the control panel, but the pipework is positioned in front of the panel. He or she has to reach between the pipes to find the correct buttons." She demonstrated the action, with the pipes blocking her way. "The buttons are close together," she continued, stretching to touch them between the pipes, "and you'll see that they need color coding to help with identification, mostly red and green, but also black and yellow."

John nodded, noticing how far she had to reach through the pipes to press the buttons.

"With our old risk assessment process, the hazard that the assessor identified was electrocution. There's even a warning sign on the front of the panel, so that's straightforward. The key point here is that the risk assessment took place without any workers present, so the assessor had to imagine how tasks in this space would be completed."

Sarah stood in front of the panel and pressed her hands against the pipes. "But what if we add a worker to the assessment? What if we pay attention to the task, not just the obvious hazard? Find out what the worker really has to do, and talk to them about the challenges they face?" She reached through the pipes again. "That's when error traps become visible."

John nodded. "Those pipes are a big problem. Can you see which buttons you're pressing?"

"My view is definitely blocked by the pipes," Sarah agreed.

John shook his head. "So ... you might try to make a small adjustment to the machinery, and end up shutting the system down by mistake, or worse!, all because you couldn't easily pick out the correct button."

"That's right, and there's also the issue of time. When we looked more closely at the work as-done for this task, we found that some workers were using the equipment during the night shift, when they could be expected to be tired. All these factors together increased the likelihood of failure, but the standard risk assessment didn't pick up these extra risks and error traps at all."

Sarah pulled a sheet of paper from her clipboard and showed it to John.

"I use this example for training," she explained. "Here's a comparison between the risk assessment for this panel from when we were looking for obvious hazards, and an assessment based on uncovering error traps."

Hazards	**Error Traps**
Electrocution.	The pipes prevent the worker from getting closer to the panel.
Hand entrapment.	The color of the buttons, if the worker is color-blind, they will not see the difference between the red and green button colors.
Temperature and burn.	Size of button labels, difficult to read from a distance.
Hearing damage.	Time, it is 3.30am, and the worker is tired and sleepy.
Pressure release.	Worker is alone.

	The work instruction he was given is out of date and does not describe how the task should be actually done.
	The radio allowing contact with the control room is not working.

John sighed. "It looks as if the standard risk assessments were overlooking a lot of critical context."

"That's right," Sarah said, glancing at her watch. "And I'll show you another example. Follow me."

Sarah led the way through a set of double doors into an empty work area.

"While it's quiet and everyone's out on their break, take a look at this," she said, pointing out the crane above their heads. "This is used by two workers, one controlling the lifting equipment, and the other interacting with the box being lifted." She pulled another sheet from her clipboard. "Like this," she said, holding out a photo to John.

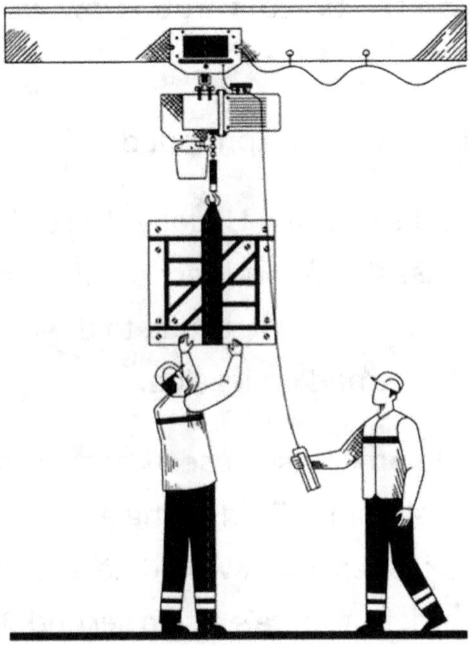

John felt his chest tighten. He could immediately see the similarities with the accidents on both his sites.

"What would you identify as the primary hazard here?" Sarah asked.

John inspected the equipment in front of him and thought about Harry's injuries. "I'd say the risk of a crush injury if the load swings or drops," he said.

"Right," Sarah replied. "But what you don't see, standing here looking at the equipment and not the workers, is the barcodes. The barcodes are typically on the side or the top of the boxes. But very rarely, around one box in five hundred, the barcode is on the bottom. You know when we talked about variability?"

John nodded, and Sarah continued.

"This is a good example. Usually there is no problem, but the risk increases when workers need to scan a barcode underneath the box. Even worse, due to the fragile contents, the box cannot be turned on its side.

"When this equipment is in use, there's a crane operator at risk from a crush injury if anything goes wrong. The old version of our risk assessment would have picked this up as the primary hazard. But there's also a second worker." Sarah stood under the empty crane and reached up to demonstrate. "The second worker's job is to scan the barcode. When we talked to the workers about error traps, we learned that when the barcode is on the bottom, they have to handle and steady the load above head height, which *increases* the risk of it falling. There could be injuries to both workers, and the work they are doing *makes those injuries more likely*."

John was surprised. "That's something that wouldn't be caught by just noting the hazard of a 'dropped load'."

"Exactly," Sarah confirmed. "And there are more complications." She counted them off on her fingers. "For instance, as the box can't be lifted on its side due to the fragile items inside, the scan might be missed. Without scanning the barcode, the box won't be accepted by the courier system, so the dispatch would be delayed. This would probably mean additional costs, a backlog of boxes in the warehouse, and pressure on the workers to process the boxes more quickly. These factors create an error trap that *significantly* increases the risks, but they were invisible when the assessment only focused on the primary hazard.

"And look how much the original risk assessment missed," added Sarah, showing John another assessment from her clipboard.

Hazards	**Error Traps**
Dropped object, crush injury.	The label with the bar code that needs to be scanned is located on the bottom of the box.
Worker could be struck by the load.	The load cannot be put on its side on the ground or in the sling due to fragile items inside.
	Without scanning the code, the load will not be accepted by the courier system, leading to costs and delays.

John shook his head. "It's amazing how much gets overlooked when you're only looking for hazards."

Sarah smiled, realizing she was getting through to him.

"Let me show you one more change we've made as a result of looking for error traps."

She led the way through a loading dock and out of the building. Round the corner from the loading dock, she stopped next to a line of chemical tanks.

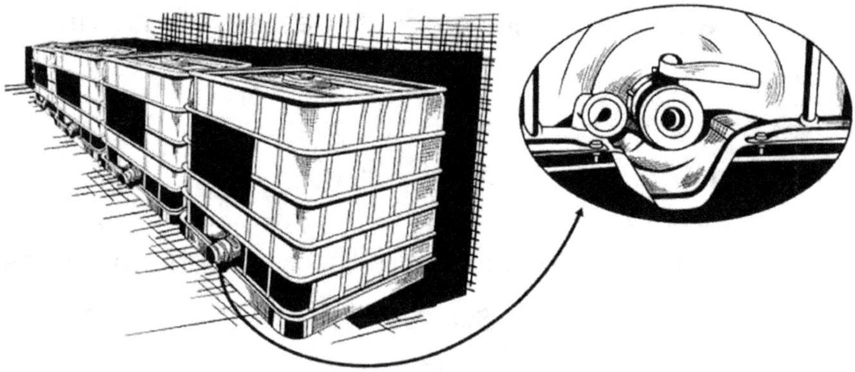

"There are some obvious hazards here," she said, pointing at the tanks.

John nodded and gave his instant assessment. "Chemical leaks, and potential chemical burns to the operators."

"Right," Sarah replied. "But what about the error traps?"

John stared at the tanks, trying to imagine how the workers would interact with the equipment. "You're right," he said, shaking his head. "Without seeing the tanks in use, it's really hard to assess what the problems might be."

Sarah stepped over to one of the tanks and pointed at the hose connection. "One major issue was that the operator could disconnect the hose, here, before closing the valves. There was no visual indication to show whether the valve was open or closed."

John's eyes widened. "That's a serious problem! That would directly lead to the chemical leak, and chemical burns!"

"It is," she agreed. "And without noticing and addressing these error traps, our original risk assessment was incomplete. The design flaw that allowed the hose to be disconnected while the valve was open wasn't something that a basic hazard analysis would catch."

"So, what did you do?" John asked.

"We started by recognizing that our risk assessments were too focused on visible, static hazards. We needed to incorporate *dynamic* risk assessment, which takes into account the changing conditions and the real-time decision-making that workers face every day."

"Working with the workers, and understanding the choices they make," John reflected. "That makes a lot of sense."

Sarah nodded. “We also redesigned our processes, assessment templates, and competencies, to better reflect the reality of the work as-done. We made sure our assessments focused on the error traps, as well as on the visible hazards.”

Sarah unclipped another sheet from her clipboard. “Here’s our updated risk assessment form in use. As you pointed out, we’ve made sure that identifying error traps is a core part of the process.”

“And this works?”

Sarah nodded. “It’s definitely an improvement on our old system.”

“But how did you get the workers on board?” John asked.

“Well, at first people saw our new risk assessments as another pointless paperwork exercise. They were too time-consuming and inconvenient, and there was low workforce involvement.”

John nodded. “And I can see that if workers don’t trust the risk assessment process, they won’t trust other safety measures, either.”

“Absolutely,” Sarah said. “And it’s not just about safety. If managers don’t trust the risk assessment process, they are likely to be constantly anxious about whether they’ve done enough to prevent accidents. They worry about having to

Sequence of Job Steps	Potential Hazards	Hazard Controls	Error Traps	Error Trap Optimizations
Step 3. Close the valves and disconnect the hose.	Chemical leak. Chemical burns.	Chemical Awareness Training. Spill Response Procedure.	Possible Error: Operator disconnects the hose before closing the valves. Causes: No visual indication if valve is closed/ open. Technician may assume it is closed when it is open. Design allows disconnection of the hose with an open valve on the tank.	Redesign the connection to make it impossible to disconnect the hose with the valves open. Add visual indication to show when the valve is open/closed.

communicate a death or a life-changing injury to a family member, or about losing contracts due to poor HSE performance."

John was quiet for a moment, processing everything Sarah had shown him.

"So, what's at stake is more than just safety," he said. "It's trust, peace of mind, even the survival of the business."

"That's right," Sarah said, leading John back toward the office building. "Redesigning our risk assessments to capture not just hazards, but also error traps and context, that's allowed us to make the workplace safer and develop trust across the organization."

John looked at Sarah with deepened respect. "This has been a real eye-opener. Thank you for explaining all of this."

"I'm glad it's been helpful, John," she said as they reached the security desk. "Remember, the key is to focus on learning and improvement, rather than compliance and blame. It will take time, but if you can change all that, you'll be heading in the right direction."

Chapter 16

The first insights from the team

John knew the next step in rolling out the Learning From Normal Work initiative was to build capability within his division.

He and his Division Management Team had been leading by example, working through the Learning From Normal Work Competency Pathway. The group began holding regular virtual sessions, tackling one topic at a time. For each topic, they started with the basics and moved to practical activities.

After two months of steady progress, John called an online team debrief to discuss how things were going, and for the team members to share their examples of practical applications.

Ahmed, from Oman, was the first to present. He took over from John as presenter and cleared his throat. "I've been doing a lot of reading on my own recently," he explained, "but this training is different. Less academic theory, more hands-on tools and practical stuff."

He paused, checking his notes, then continued, "I want to share an exercise from the module on design. I've got an engineering background, but we never learned about human factors and usability in my degree.

"Recently, we ordered a new machine from a leading provider. It's been in use for about three weeks now. Using the template from the training, I talked to the operator, and we found design issues that no one had noticed before."

John leaned forward, intrigued. "What kind of issues, Ahmed?"

Ahmed glanced at his notes. "Well, for starters, the pressure gauge is on a different side of the machine than the controls. The operator has to control the machine and monitor the gauge at the same time, but the design makes that impossible. It's either a two-person job, or the operator has to fix the control levers with a cable to run around and check the pressure. It's ridiculous when you think about it."

The meeting was silent as Ahmed continued. "So, prompted by the checklist from the training, I checked our design requirements for the equipment we purchase. I was very surprised to find that we hadn't specified anything about usability, human factors, or user involvement. We hadn't told the manufacturer that we needed a machine for a single worker to operate." He shrugged. "Why aren't we doing this? Why aren't we thinking about human factors in advance?"

John nodded. "That's a really good question, Ahmed."

"I'm convinced that if we address this," Ahmed continued, "and include usability requirements into our global standard, we could drastically improve the quality of the equipment we buy."

John thanked Ahmed and made a note of his comments.

Next to present was Amy from Australia. She appeared on screen, her expression serious.

"I decided to share what I learned from the module on procedures," she began. "We often assume that rules are always available, workable, and correct. When non-compliance happens, we blame the individual for not following the rules. But this exercise opened my eyes."

She paused for a moment before continuing. "I went to the shop floor and spoke to an operator handling our pressure pumps. We did a quick screening, going through various problems with procedures, and asking which procedures have these problems.

"In just ten minutes, we identified four problematic procedures. One of them was physically impossible to follow because, like Ahmed's machinery, the equipment layout didn't match what the procedure required." Amy shook her head, clearly frustrated. "We picked one of those procedures and went through it step by step, focusing on the critical tasks. Out of twenty-seven steps, we prioritized eight as critical. Four of those eight steps were not really possible to execute. We found incorrect values, inconsistent pressure units, for example, and one step was meant for a different type of equipment altogether."

The team members nodded in understanding as Amy continued. "When I spoke to the engineer responsible for writing these procedures, she told me they have about eight hundred operational procedures, all written by engineers in the central office. There's no involvement from the people doing the work, and no process for updating or monitoring their use. If we could update this process, we could significantly reduce non-compliance and make the procedures actually helpful for the people on the ground."

Hudson from the US was next. "I'd like to share what I found in the module on predicting human error. Like the others have said, we usually think errors happen because people aren't paying attention or following the rules. We rely on training, motivation, and punishment to fix the problem.

But after going through this exercise, I see that we often set people up for failure."

Hudson looked at his notes, then back at the meeting. "I reviewed a routine maintenance task on one of our machines. Using the error-prediction keywords from the training, we identified potential mistakes that no other process had shown us before.

"For example, the bolts need to be unscrewed and screwed back in a specific order because of how the cover is designed. But the numbering on the panel and the drawing didn't match. The sequence on the panel was 1,3,2,4,6,5, and on the drawing, it was 1,2,3,4,5,6. This led to mistakes that caused vibration, damaging the parts, and leading to quality costs and delays. In the past, they had blamed the operator for not paying attention and hadn't been able to address the issue."

Hudson's frustration was evident. "The maintenance engineer told me these instructions came from the equipment manufacturer. We agreed that if we could work with the manufacturer, we could significantly reduce the risk of this type of failure. I was shocked that we could find things like this without waiting for an accident to happen."

As Hudson finished, two other team members shared their experiences, but not everything went smoothly.

Yichen from China voiced his concerns. "This all sounds good, but it's too time-consuming. Arranging discussions, going to the site, having these conversations, it's not sustainable. You can't roll this out at scale."

John thanked Yichen for his honesty and asked the team to comment. "What do you all think?"

Kate, another team member, spoke up. "Yichen, I get what you're saying. But I don't think we need the entire organization to go this deep. We're going through this in-depth program to review our highest-risk activities, and to build our own skills so we can coach others and help them implement the new techniques. Right, John?"

John nodded in agreement. "We're aiming for a change in the culture of the company, not looking for a tick-box exercise. Learning From Normal Work is an investment, and it only makes sense if we put the effort to discover things we haven't seen before, and address risks that we didn't know existed. If we're not achieving that, then the program doesn't make sense."

Yichen didn't seem entirely convinced. "Instead of us going this deep and spending this time, we should just send an email to the technicians and ask them what their challenges are."

"I'm glad you mentioned that, Yichen," said Hudson. "I actually tried that. I asked exactly that question, but it didn't work as intended. Just a few people replied, and the things they shared were either too vague, or not related to this process. Some mentioned too low salaries or wanting better food in the canteen. And I'm not dismissing these needs, HR should look into them, but there is no way these answers would give me insight into error traps and hidden hazards in a way I was able to uncover through a conversation."

"That's good to know, Hudson," said John. "We want to learn from what works and what doesn't work."

The next presentation came from Julio in Costa Rica. "I ran into a problem with the module on Learning Teams," he explained. "I gathered three workers from an assembly line, and when they started walking me through the process, I realized something really important.

"They were telling me things that other departments already knew. For example, they talked about issues with the design of the machinery, like the ones Ahmed and Amy mentioned, but it turned out that the engineering department had already been told about the problems. I had to talk to the workers to find this out."

Julio shrugged. "Surely we should be sharing these issues across all our departments! I can't know what no one tells me. And there were other things as well, problems that the facilities department were dealing with on their own. It shouldn't be up to the workers to make sure everyone knows about the issues they face."

"That's very interesting feedback," John replied. "Thank you, Julio. I'm sure you're not the only one in this position. I wonder what we can do to fix the problem."

Nadiya, the project manager, spoke up. "What if we ran our Learning Teams including people from different departments? Obviously talk to the workers, but why not bring someone from engineering along, too? And someone from planning, facilities, every department that might be involved with the process we're looking at."

John nodded. "That's a great idea. Cross-departmental Learning Teams." He made a note of the suggestion. "That's something I'd be very happy to try out."

The team continued to discuss the challenges and benefits of the training program. John hoped that he could keep his team engaged, in spite of the hard work ahead.

Chapter 17

A complex learning team

John realized it was time to step up his game. Inspired by the debrief, along with his conversation with Sarah, and by studying the work of experts on complex systems failures, John was determined to understand how these theories could be applied to his own operations.

With the help of a trusted expert, he set up a two-day Complex Learning Team session to examine the risks associated with complex lifting operations. The two recent accidents on his sites involved multi-ton lifting equipment that required precise and coordinated handling, multi-axis rotations, and coordination between multiple teams. He wanted to make sure he fully understood the factors that had contributed to the accidents.

Previous Learning Team sessions had only involved front-line workers, but this time John decided to test Nadiya's suggestion. He wanted everyone involved.

On the day of the session, representatives from various departments gathered to take part. John welcomed crane operators, engineers, and staff from logistics, procurement, HSE, and facilities.

John opened the session. "We're here not just to talk about what makes your job hard, but to find out how we all contribute to each other's challenges and successes."

As the facilitators began their work, the participants started to talk about their experiences and frustrations. John listened intently as each department explained their part in the complex process of a dangerous lifting operation.

"Important load information is missing from the documentation," one engineer pointed out. "That's because the software we use doesn't have the necessary feature enabled. It was deemed too costly during procurement."

Procurement spoke up. "We were never informed that this feature was critical. Our decisions were based on the budgets and specifications we were given."

Training issues also came to light. "Our operators are trained, sure, but not on the detailed skills they truly need for these kinds of lifts," said a crane operator.

"Our training budget is tight, and we thought general training would suffice," an HSE advisor responded.

John could see the pattern emerging, as Julio had reported. Divided decision-making was leading to gaps in skills, resources and information. As the session progressed, more than one hundred error traps were identified, and thirty were prioritized for immediate action.

But the deeper insight came from recognizing that these weren't isolated operational failures, but symptoms of a fragmented organization.

"This isn't just about a crane operator making a mistake, or a piece of equipment failing," John reflected aloud to the group. "It's about how our entire system is designed and operates. Each stage of this process impacts other teams and other activities, and right now, we're not working together."

••●••

John arrived home feeling overwhelmed. The Complex Learning Team conversations had reminded him of the size and complexity of the task ahead.

Over dinner with Melissa and Erica, he couldn't stop thinking about the progress his teams were making and wondering whether he was doing enough to support them. He didn't want them to lose interest in the program, and he certainly didn't want to set them up to fail. Most importantly, he needed to make sure representatives from different departments understood the need to work together.

He needed to overcome his concerns. Watching his wife and daughter talk about their plans for the weekend, he realized he could step back from some of the day-to-day issues. He knew Nadiya was capable of running the project, just as Melissa and Erica were competently making plans involving the three of them.

Stepping back would give him time to inspire his management team to work together with all the departments in his division. Training champions at various levels in the organization meant that he didn't have to run everything himself. If he could take the lead on bringing departments together, he felt sure that Nadiya and the champions could do the rest.

"Are you OK, John?" Melissa asked. "You're really quiet this evening."

Erica rolled her eyes. "What's new?" she mumbled.

John rested his elbows on the table and smiled at his wife and daughter. "I think I am," he said. "I think I know how to make this project work."

Chapter 18

The data review

As the data from the initial phase of the Learning From Normal Work program started rolling in, John found himself anxious and eager to see the results. He needed to understand how the sites involved in the first stage of the roll-out were progressing. More importantly, he needed solid evidence to present to the CEO, proof that this initiative was making a real difference.

John called his project manager, Nadiya, into his office. “I need you to compile all the data we’ve been collecting,” he said, his tone serious. “I want to see the progress we’ve made with training, and I really need to see the impact all this effort is having on safety. This hasn’t been easy or cheap, and we’ll need a clear demonstration of what we’re getting for our investment.”

Nadiya nodded. "I'll get the data together, and I'll send out a survey to the leadership teams involved in the first phase. I'll arrange a few calls to dive deeper into their experiences."

A few days later, Nadiya returned with the compiled data.

"It's good news, John," she began. "The feedback is mostly positive. About seventy-five percent of respondents are seeing safety improvements, but twenty-five percent mentioned opportunities to improve the roll-out. I think we'll need to make adjustments to fit the local needs on some sites, but no one has rejected the program or withdrawn, so that's an improvement over the first pilot."

John nodded, but he wasn't completely satisfied. "That is good news," he said, "but I need more than just feedback. I want to know if we're actually reducing risk. What are we finding? How are we preventing accidents?"

As John read through the data, he noticed that different sites had set their own targets for assessing training and safety improvements. He had suggested this as a way to avoid distorting the data with unfair site-to-site comparisons. He was pleased to see the site teams engaging with the roll-out.

He turned his focus to leadership behaviors and engagement. The frequency of their involvement varied between sites, but the engagement overall was clear and significant.

One section of the data caught his eye. “These Learning From Normal Work discussion cards,” John muttered to himself. He scanned the results, noticing that conversations using the discussion cards were producing useful insights. He was relieved to see that this approach was more successful in uncovering issues than previous investigations that focused narrowly on compliance.

Nadiya stepped in, sensing John’s curiosity. “Look at this,” she said, pointing to a specific example. “Training was being given for a job, but the reality of how the job needed to be done was entirely different. And here, tools required a lot of physical effort, leading to dangerous workarounds like ‘cheater bars’. The workers were adding long pieces of pipe to their spanners to make them easier to turn. In another case, equipment controls didn’t provide feedback, making it difficult for operators to understand what was happening.”

John nodded, impressed. “These are the kind of insights we were missing before. This is great news, it’s not just about ticking boxes anymore. People are taking this seriously.”

He then moved on to review the Walk-Through Talk-Throughs. As he compared the quality of these sessions, he remembered his own first experience, and how difficult it had been to get the process right. He noticed the same mistakes happening again, but something else stood out.

"When I compare the walk-throughs from teams supported by the champions to those that weren't," John said, his voice rising with excitement, "the difference is obvious.

"The champions are coaching people to achieve better results. We need to address this moving forward. I'll send out an email to thank those champions for their hard work."

But as he dug deeper into the Walk-Through Talk-Through reports, John's excitement turned to concern. These sessions were uncovering important activities and steps that had never been identified before. While the new approach was a significant improvement, it also meant their previous understanding of risks had been dangerously incomplete.

"The procedural steps are better now," John noted, "and we're also seeing redesigns of tasks, improvements in engineering controls, and better alignment between teams. This is the kind of depth we've never had before."

But even with all this progress, John knew the CEO would ask the inevitable question, "So what? How is this helping us prevent accidents?"

John leaned back in his chair, knowing he couldn't rely on the injury statistics. Most of the sites had low numbers of reported injuries, both before and after the roll-out, making that metric almost useless. Instead, he focused on risk reduction.

They had taken the original risk assessments, reviewed the activities through walk-throughs and Learning Teams, and then recalculated the risk after identifying error traps and problems.

What they discovered was startling. "When we account for these error traps, the actual risk level is much higher than we thought," John said, the realization sinking in. "Our previous risk assessments were missing key aspects of operational risk. We focused on hazard controls, but not on the error traps and hidden issues."

Nadiya looked at him, concern showing on her face. "John, this means we've been operating with a false sense of security."

John nodded. "Exactly. We've been underestimating the risks. This new insight isn't just about making small improvements; it's about correcting a fundamental misunderstanding of our operational safety."

Nadiya looked thoughtful. "So how do we use this information to convince the CEO? He's not going to like seeing our risk levels increasing."

John turned back to the calculations. "We can't present the findings without explaining all the work we've done. We need to make sure he understands that these risks have always been there. Operating without knowing about them is far more dangerous than pretending they don't exist."

"So he's waiting for a presentation? He wants to know what we've found?"

John nodded. "As soon as possible."

Nadiya gathered the report pages from the table in front of her and opened her laptop. "Let's get started."

Chapter 19

The end: presenting the results

A few days later John met Nadiya and Karen in the conference room to prepare for their presentation.

"Are you ready?" Karen asked. "Do you think you can convince him?"

John nodded. "I hope so. We've put a lot of work into this. We want him to understand everything we've been working on, and everything we've learnt. If he can see what we've discovered, I'm sure he'll understand the benefits of taking a new approach to HSE."

When the CEO arrived, John welcomed him and began by examining the shortcomings of the investigation into Harry's preventable injury.

"It was a shock to be faced with such a serious accident, after two years of a perfect safety record," John explained. "But placing all the blame on an experienced crane operator like Dan? That didn't feel right to me. I felt sure there was more to the accident than his alleged incompetence."

The CEO nodded thoughtfully. "And after the second accident on your watch, we blamed you for incompetence."

"That's right," John said. "In a way, the second accident, Lisa's accident, was even more shocking than Harry's. We thought we had conducted a strong investigation, and we'd brought in additional training for the crane operators. We thought we'd solved the problem, and then almost the same thing happened again. And this time, it was my reputation being questioned." He smiled. "I could certainly see why the workers were concerned. Just like me, they were worried about being blamed for things outside their control."

"And that's when you brought the idea about systemic issues to me," the CEO said.

John nodded. "I'd been reading up on better ways to conduct accident investigations, and I found the idea of Learning From Normal Work really intriguing."

Karen stepped in. "I'd been looking into Learning From Normal Work, and Human and Organizational Performance, as well. When John brought up the idea, I thought it could bring a new perspective to our safety management strategy. John set up the workshop with the senior leadership team so we could learn from an expert how to apply the ideas."

The CEO frowned. "As I recall, the team's reaction wasn't great, was it?" he asked.

Karen responded. "It's a big subject to explain in a short workshop," she said. "I don't think we had a proper chance to understand the details. Most people seemed to think it would add more time and paperwork to our safety systems, when in fact it could transform them entirely."

"When I visited workers on our sites," John explained, "and sat down with them to talk about their concerns, I could see they were really worried. They'd seen that after Harry's accident, Dan had been disciplined, but they all knew how easily an accident like that could have happened to anyone. They knew about problems with machinery and procedures that we'd missed in our investigations. It was impacting their work and their morale, and productivity was dropping.

"It was as if they'd given up," John said. "As if anything they did might result in punishment."

"So, this is when you ran your pilot study," the CEO waited for John to continue.

"Almost." John replied. "I had to learn a lot about the techniques and ideas first. Learning From Normal Work requires trust, from the workers and the managers. They were used to seeing me when something bad happened, treating them as if they'd done something wrong, and trying to catch them out in a dangerous activity. To them I was the investigator. If I decided they had acted dangerously, they could be disciplined, so they didn't want to talk to me about problems in their workshops. They were used to keeping quiet and protecting each other."

"So how did you fix that?" asked the CEO.

"I had to learn how to ask questions, and how to listen," John said. "I had to keep going back, keep having conversations, and keep showing them that they could trust me. I fixed some of the obvious issues they raised, and when they realized I was actually there to listen, and not trick them into getting into trouble, they started to tell me what was really going on.

"That's when we ran the pilots, when I was confident that I had understood how to put the techniques of Learning From Normal Work into practice."

"And they went well, I think?" The CEO looked expectantly at John.

"We had mixed results," John said. "We made the mistake of running the first pilot at one of our worst-performing sites. We weren't ready for problems with the pilot, and they didn't have the resources to fully commit to making the changes in their processes. We couldn't support them, and they found everything we asked them to do overwhelming. They ended up pulling out of the pilot."

"That seems like a waste of time," the CEO commented.

"Yes and no," Karen said. "It showed us how hard it could be to convince and support an already struggling site to make all these changes."

John nodded and continued his explanation. "We also realized that we needed to offer more training to the other pilot sites, to make sure they properly understood the techniques we were trying to introduce. We ran workshops to prepare the management teams, and I made frequent visits to the sites to top up the training and answer questions.

"For the third pilot, we provided a much more structured roll-out and training program," John explained. "At the same time, we put together a set of measurements we could use to assess the success of the pilots."

"And I heard great things," said the CEO. "Those site managers were singing your praises. I saw your report, too. All those cost savings, I had no idea it would be so expensive to react to an accident, but so cost-effective to run this training instead!"

"It's a huge saving," Karen agreed.

"So," the CEO said. "Your division roll-out. Tell me how that's going."

"It's going well," John replied. "We learnt from our pilot sites, and we're providing in-depth external training for our management teams, Division-level, site-level, and departmental managers, as well as HSE champions at all levels. We're running planning workshops to make sure we are meeting the support needs of the entire Division, and we're providing regular updates and communication between teams. We've invited honest feedback, and made sure the management structures are in place to make the roll-out a success."

"It sounds as if you've built on what you found on the pilot sites." The CEO nodded approvingly. "And what about the results? Are the new techniques working?"

Nadiya answered his question. "Feedback from the site managers has been overwhelmingly positive. More than three-quarters of our sites are reporting improvements in their safety numbers. We're also getting some very useful feedback, allowing us to adapt the training and implementation on a site-by-site basis.

"Through non-judgmental conversations with workers and supervisors, we've picked up plenty of safety concerns that were not showing up in our traditional risk assessments," Nadiya continued. "Rather than hiding their workarounds, workers are now happy to share the issues with us. We've demonstrated that we understand the problems they face in their day-to-day work, and instead of blaming them for solving problems in unauthorized ways, we'll step in to fix the issues and keep everyone safe."

"That sounds like an improvement," the CEO said. "But does it help? Does it stop accidents from happening?"

"We believe so," said John. "We can show that most sites have reported fewer instances of unsafe behavior, and that we are fixing problems before they become dangerous. For instance, we have replaced all the faulty remote controls for our cranes. Workers no longer have to open up the controllers and adjust the electronics in order to do their jobs. Before, they were hiding the faults, and their actions, because they knew they would get into trouble. Now they trust us to fix the problem instead of blaming them for equipment failures.

"We've also made it easier for departments to work together," John continued. "The crane controllers hadn't been replaced, because the workers were told by procurement that there was no money for repairs. They were told to solve the problem themselves. Instead of reporting this as a safety issue, they worked out how to open up the units, risking further damage and electrocution. Now, procurement, safety, engineering, and anyone else involved work together to address problems like these. No one is afraid to speak up, because we all recognize the importance of working together to keep workers safe."

The CEO nodded. "Very impressive, John. Does this mean that our risks have reduced?"

John took a moment to gather his thoughts for the hardest part of the presentation.

"Not exactly," he said. "But allow me to explain.

"People are speaking up now and telling us about risks we didn't know about before. We've changed our approach to risk assessments, too. Instead of simply looking for the obvious hazards, we also look for the error traps and hidden problems that might force a worker to choose an unsafe action."

John paused, but the CEO gestured for him to continue.

"Our improved assessments are showing higher levels of risk after the roll-out, but I would encourage you to see this as a good thing.

"Those risks existed before. They've always been there; we just couldn't see them. So, while our risk levels appear to have increased, we're actually just seeing the true level of risk for the first time."

The CEO frowned. "OK," he said. "I see your point. So how does this make us safer?"

"The main thing to understand is that this program hasn't been about small improvements here and there. What we've been doing is changing the entire culture around safety, behavior, and blame," John said. "Now that workers trust that we're not going to blame them for problems with equipment or procedures, they are happy to report safety issues, and this gives us the ability to minimize or fix the risks.

"We've brought different departments together to plan solutions for the problems we find, so we can fix them quickly. We can find the budget for urgent repairs or changes, because we all understand how important it is to do so. Departments are working together on safety, rather than constantly passing on problems to someone else to solve.

"So, the numbers might look worse," John concluded, "but the change in culture means that our workers will be safer, now that we understand all the risks."

"And let's not forget the cost savings," Karen said. "Understanding and acting on all these risks is dramatically cheaper than dealing with serious accident."

John held his breath and waited for the CEO's response.

The CEO leaned forward in his chair. "This is an impressive amount of work, John. It's a lot to take in, but I can see you've considered this approach from every angle. I have some questions, but I can clearly see the benefits of a more open reporting system. You've opened my eyes to the advantages of basing HSE management firmly in the reality of day-to-day work.

"And I want to personally apologize for the warning from HR. I can see now that the accidents weren't due to anyone's incompetence. Before all this, I couldn't see the bigger picture, and I needed someone to blame." He gave John a smile. "I'm very grateful that you put all this work in, and I'm glad you didn't give up."

John took a deep breath, a feeling of relief rushing over him.

"I've seen the results of my other pilot, too. That Division made some progress, but your program dug much deeper into the issues and challenged how we all work together as a business. You involved all your departments, instead of just focusing on the front-line employees."

The CEO smiled.

"Congratulations, John. I think this is the way forward, and I want to invest in a better, safer, future for this company. You should roll out these changes across the business."

John couldn't help smiling. "Thank you," he said. "I can't wait to get started!"

"We also have a customer audit soon," the CEO continued. "Glacier Manufacturing have some projects coming up that we'd like to work on, and we've put in a bid for the contracts. Small projects to begin with, but there's a major project coming down the pipe, and we want to convince them to choose our bid. This approach could tip the balance in our favor when they start looking into our HSE record."

Karen sat forward and rested her arms on the table. "It's a great opportunity, John. A major contract with Glacier will keep the shareholders happy and do wonders for our reputation as a trustworthy company to work with."

The CEO nodded. "This could be just what we need to convince them to accept our bid. Can you turn all of this into a report for their audit? It would be good to be ready when they start their research."

"Of course," John said, feeling a rush of pride in everything he and his team had worked on. "I'm happy to share our findings with everyone who can benefit from the work we've done."

Chapter 20

The new beginning: two years later

John straightened his smartest jacket and gave Melissa a quick smile. She looked stunning tonight in her new dress, John thought, as she smiled back. They made their way across the crowded room to the table set aside for the Avriox Industries team. Karen and Nadiya stood up to greet them, and before long Melissa was sitting next to Karen, deep in conversation. Nadiya took the chance to speak to John.

"Nervous?" She asked, and John nodded, patting the pocket where he'd carefully placed the notes for his speech. "Tonight is a big deal. You're allowed to be."

John looked round at the grand hotel ballroom, the large round tables set up with white linen and sparkling place settings, and representatives from large multi-national companies dressed up for an evening of celebration. He couldn't help thinking about the last award party he'd attended, and the award he'd been given for his spotless safety record, and about the morning after, when he'd been called to the factory by Mark, and found Harry so badly injured.

"I can't believe how much we've learnt since last time I collected an award," he said, and Nadiya nodded. "Thinking about what I thought I knew then, and what we know now ..." He shook his head. "I don't know how we avoided an accident for so long without properly understanding the risks we were working with."

Nadiya smiled. "You've really made a difference, John," she said. "To the whole company. You've brought people together, from the factory floor to the CEO. There's so much more trust, now. You've taught people to recognize each other's expertise, and to share their concerns without the fear of being blamed. You should be proud."

Karen nudged his elbow. "She's right, you know. You've managed to lead a shift in the culture of the company. That's no small achievement."

John's thoughts were interrupted as a waiter brought bottles of wine to the table and filled their glasses.

"Oh! I haven't had a chance to tell you." Karen leaned her elbows on the table when the waiter moved on. "Glacier Manufacturing signed the contract! They said our approach to HSE impressed them enough to make us the front-runner for the major project bid. The biggest contract is ours!"

Nadiya and Melissa raised their glasses. "Cheers, John! That's great news."

As everyone at the table drank a toast, John's phone lit up with a text from Sarah. *Congratulations on the award! I hope you're having a great time. You deserve it.*

He texted back. *Thank you! I couldn't have done it without your help.*

••●••

John's speech passed in a blur. He outlined the steps he had taken with his team to implement Learning From Normal Work across the company, explained the learning process, including the failures as well as the successes, and thanked everyone who had played a part in researching and rolling out this new approach to HSE management.

The Chairman of the industry body hosting the event shook John's hand and presented him with the award for Outstanding Safety Leadership. As the audience clapped, he walked back to his seat with the heavy glass award in his hands. He felt the weight of the mistakes he had made before, and the confidence in his ability to react differently now, to risks, to reported problems, and to future accidents.

The CEO stood as John approached their table and shook his hand. He spoke quietly. "Well done, John. I couldn't see the benefit of all these changes when you started, but you convinced me. We're better off now so that we can assess our risks more accurately. And after the Glacier contract's success, I think this is going to be another unique selling point for the company. The competition will be racing to catch up!

"Have I ever told you what the company name means?" John shook his head, and the CEO continued. "It's from the Greek word for 'tomorrow'. Our founder wanted to push the boundaries of technology when he started the company. He wanted to use tomorrow's technology today. I think that's very appropriate, don't you? This new approach to HSE management feels like tomorrow's standard, in use today, and that's down to you and your hard work."

"Thank you," John offered another handshake. "And thank you for your support."

"Congratulations," Melissa said as he set the award down in front of her. "You must be relieved that the project is over."

Karen put a gentle hand on Melissa's arm. "I'm sorry to be the bearer of bad news, but we're going to keep working on our approach to HSE, and your husband is going to be at the heart of the process." She smiled at John. "We can't let him stop now! We need to keep improving and sharing our findings."

"Karen's right." John settled into his chair and picked up his wine glass. "This isn't the end of the project. This is just the beginning. We're going to keep getting better, and finding new ways to keep everyone talking, and sharing their best practice and their concerns."

Nadiya laughed. "Hoping for another award, John?" she asked.

John thought for a moment. "Honestly, it's not about the award," he said in a serious tone. "I accepted a safety award once before, and immediately realized I didn't deserve it. The industry recognition is nice, and this dinner is lovely, but the important stuff is happening at work on our sites. There are workers who feel they are being listened to for the first time. We're taking their concerns seriously, and we've committed to keep listening, keep learning, and keep improving. We've made our sites safer, we've stopped blaming individuals for systemic problems, and we're preventing accidents, not waiting for them to happen." He took a sip of his wine and looked around the table. "That's the reward that matters."

Karen raised her glass. "Well said, John," she declared. "And well done."

As the team drank another toast, John's phone lit up again. This time he found a text from his daughter. Melissa had sent her a photo of John receiving the award, and she'd sent it back to him with a comment.

Proud of you, Dad.

Case studies and industry efforts

John's story is based on real situations from various companies. In this section, I also want to share the real-world stories of organizations across a number of industries, highlighting their shifts to proactive learning.

ABB: HEAVY MANUFACTURING (International)

From Guiding Principles to Action By Mitch, ABB Global HOP and Investigations Manager.

The Beginning: Searching for More Than Compliance

My journey with Human and Organizational Performance (HOP) at ABB started in 2016 when a colleague, Guillaume Alzina, introduced me to the concept. At the time, I was ABB's Global Investigations Program Manager, and I was curious, but HOP was just another approach among many. I wasn't sure how it could fit within ABB's structure or if it would gain traction.

By 2019, my perspective began to change. Division HSE Manager, Mark Sage, was struggling with repeated incidents in his Division and wanted to move beyond traditional incident investigations. I saw an opportunity to evolve how we learned from normal work, so I began reaching out to industry experts to explore how HOP could apply within ABB. With Mark's support, and sponsorship from our Asia Business Manager, we engaged with an external HOP training provider to deliver a HOP and Learning Teams pilot in Asia.

The pilot started with strong leadership support, bringing together Local Division Managers and teams from across Project Management, Supply Chain, Engineering, Operations, and HSE. After an intensive five-day training, enthusiasm was high - people saw the potential for HOP to change how we learned and improved. But then, a week following the pilot, our Asia Business Manager resigned from his role.

Momentum disappeared. Over the next six months, nothing materialized. Twelve months later, we pushed for a few pilots, but enthusiasm had faded. Every time I suggested building HOP competence, I heard the same response: *"I don't know... we don't want another failed pilot, do we?"* It was frustrating, but I understood. I continued learning about HOP on my own, but without leadership support, it felt like it could be another industry fad that would pass.

The Breakthrough

While our pilot struggled, Mark had already started laying the groundwork for ABB's HSE Guiding Principles (GPs) in 2021/2022. Recognizing that incident rates had plateaued, ABB leadership engaged in internal discussions and external consultation to explore a better approach beyond compliance.

Then, in 2022, change began to feel possible. Our Global Division President attended a customer safety forum where oil and gas leaders discussed their HOP-based behavioral safety programs and the improvements they had achieved. Inspired, he started messaging Mark:

- *"Should we be doing this?"*
- *"I want to know more."*

For the first time, we had real executive engagement. This wasn't just another passing trend, customers were embedding HOP, and it was driving real results. If ABB didn't follow suit, we risked falling behind. This curiosity and executive sponsorship resulted in the formalization and launch of our HSE Guiding Principles for Resilient Operations.

The principles were officially introduced through a Global Webinar in late 2022, continuing across 2023 and 2024 to ensure every employee was touched by our Guiding Principles.

Activating the Guiding Principles: From Concept to Reality

With leadership backing, we set out to activate the Guiding Principles (GPs), ensuring they became more than just words on paper. We focused on practical actions:

- Guiding Principles Awareness Training (2022-2024): A multi-year rollout to embed the Guiding Principles across ABB's global workforce.
- **Lead With Care training workshop (mid-2023-ongoing):** A leadership development initiative designed to help our leaders across all functions understand and apply the Guiding Principles while shifting their mindset toward learning over blame. This initiative was delivered throughout 2024 to reinforce leadership commitment.
- **Safety Circles (Jan 2025 ongoing):** Quarterly Safety Circles hosted by Line Managers with their front-line workers aim to build psychologically safe workplaces where our front-line workers feel comfortable to raise their concerns about workplace related risk. Common workplace scenarios are discussed and our front-line workers feel comfortable to relate and share their "work as normal" experiences. Discussion follows among the team and their Line Manager on how to respond to these common workplace pressures to reduce workplace risk.

These programs ensured that leaders weren't just endorsing the Guiding Principles, they were actively sponsoring and participating in the journey.

Testing Learning from Normal Work: The India Incident

In May 2023, our commitment to Learning from Normal Work (LFNW) was tested. While attending an ABB Global HSE Leadership meeting in Zurich, we got word of a serious contractor fall in India, a worker had fallen 8 meters through a false ceiling and was in a coma in ICU.

Normally, an incident like this would lead to a traditional investigation, focusing on rule violations and assigning blame. But this time, we saw an opportunity to apply Learning Teams, engaging those closest to the work. I volunteered to lead the investigation, with one condition: *we wouldn't take the usual approach.* Instead, we would focus on understanding work as-done and leverage socialized discovery techniques - an adaption of Learning Teams.

Over two days, we brought together workers, planners, contractors, and emergency responders to reconstruct how work was actually happening. What we learned changed everything:

- Workers identified previously unseen risks that had been missed in formal risk assessments.
- We introduced AI-driven contractor management solutions, based on insights from those who did the work.

- We validated and reinforced emergency response planning, discovering that paramedics had saved the worker's life because they had previously rehearsed similar scenarios.

This was more than just an investigation, it was a proof point for Learning from Normal Work. The India facility has since become a best practice example for ABB, showing how engaging workers in learning, not blame, leads to real improvements in safety and operations.

Where We Are Now: Simplifying for Success

By late 2024, I stepped into the Global HOP Manager role, carrying three key lessons from the past eight years:

- Do fewer things and do them well.
- Get early wins, build credibility, and scale from there.
- Keep leadership engaged and invested along the journey.

Rather than trying to implement HOP all at once, we focused on practical, structured actions. A major step was engaging Dr Marcin to roll out his Learning from Normal Work (LFNW) Competency Pathway, ensuring that our HSE specialists could apply HOP principles in real work settings.

Looking Ahead:
The Future of HOP in ABB

We're still at the beginning of our journey, and we are okay with that. What matters is that today, we have:

- A strong foundation with ABB's HSE Guiding Principles for Resilient Operations.
- Committed leadership, actively supporting the shift through Lead With Care.
- A structured, applied approach, ensuring Learning from Normal Work is part of daily operations.
- A growing team of ambassadors, embedding these principles at every level.
- Strong partnerships with experts like Dr Marcin, helping us stay on track.

The real measure of success isn't in what we call it, it's in how people identify their role and "Play their part". We're exactly where we need to be on the path to becoming a world-class learning organization.

Mitchell Cini serves as the Global Event Learning Manager for ABB, where he also leads the Global Incident Investigations Program. A strong advocate for Human and Organizational Performance (HOP), Cini has been instrumental in integrating proactive safety methodologies across ABB's global operations. He has facilitated HOP training initiatives, including a regional pilot program in Asia-Pacific, and has shared his expertise at industry conferences such as the Australian Low Voltage Conference, where he presented on applying HOP principles to low-voltage electrical work activities.

About ABB

ABB is a global technology leader in electrification and automation, committed to enabling a more sustainable and resource-efficient future. The company's purpose is to drive progress through innovative solutions that enhance performance while reducing environmental impact. ABB's culture is anchored in four core values: **Courage**, **Care**, **Curiosity**, and **Collaboration**. These values guide the company's actions and interactions, fostering an environment where employees are empowered to lead, innovate, and work together towards shared goals.

DOMTAR: PULP AND PAPER (USA)

Domtar's Journey to proactive learning

Since 2013, Domtar Corporation (Domtar) , a leading provider of pulp, packaging, and paper-related products across North America, has been on a journey to evolve how it approaches safety, performance, and learning. While the company has long prioritized safe operations, a pivotal shift began when traditional reactive approaches no longer offered the insights needed to sustain improvements.

Laying the Foundation with Human Performance

In 2014, Domtar's safety department initiated a series of Human Performance Improvement (HPI) training sessions aligned with the Department of Energy's HPI guide documents. Over 500 employees across various facilities were introduced to foundational principles through this effort. Initially, the focus was on improving investigations and enhancing reactive learning. HPI was also integrated in many other functional departments across the company. These steps helped strengthen how Domtar responded to incidents and learned from them.

But success brought a new challenge.

As incident rates declined across the business, leaders recognized a critical gap: *How can we continue learning and improving when fewer things go wrong? How can we move to a more refreshed approach of* Human and Organizational Performance (*HOP), viewed as not just a safety initiative or investigative tool, but a foundation of how we do business?*

The Shift to Proactive Learning

Domtar began exploring more proactive methods and even developed some of its own approaches and tools to drive identification of common practices and human elements. The goal was to find ways to learn from everyday work, not just from failures, and move to a more proactive approach to HOP. As Domtar looked to expand its learnings and toolbox, they began to look at consultants who could bring more robust education, and more modern approaches and tools. Domtar found many providers of HPI and HOP training focused heavily on concepts and philosophy. What was missing were **hands-on tools** that supervisors and managers could apply directly in the field.

To move from concept to application, Domtar realized it needed not only practical tools but also strong internal buy-in to support wide-scale adoption.

Building Awareness and Buy-In

The turning point came in 2023 when Dr Marcin Nazaruk presented at the Pulp and Paper Industry's annual conference, offering attendees, including many Domtar employees, their first look at proactive learning in action.

Following the conference, Marcin was invited to present internally to executive leaders, mill managers and union representatives. The message was clear: this approach is not a replacement of existing systems, but an **evolution**, building on what already works and taking it further.

Domtar then launched a pilot program across four selected sites.

From Pilots to Practice

At the pilot sites, Marcin delivered Mindset Shifter workshops for leadership teams and conducted one-day practical training sessions for frontline supervisors. Additionally, a group of employees from various mills joined the Competency Pathway, building deeper skills to act as internal coaches and champions of the new approach, followed by a day-long workshop for the executive team.

As the tools were applied, sites began uncovering new challenges and opportunities that existing systems had missed. These insights came directly from engaging with frontline workers, insights that could only be revealed by listening to those closest to the work.

Mills began using eLearning modules to ensure that front-line workers understood the core concepts and the rationale behind the new tools and mindset.

Union Support and Shared Commitment

A key factor in the early rollout's success was the involvement of union representatives. They attended the workshops, offered feedback, and became enthusiastic advocates of the approach. Their support helped increase trust and alignment across levels of the organization.

Looking Ahead

While the journey is just beginning, Domtar is making strides toward embedding this proactive, learning-driven mindset across the enterprise. The experience so far has reaffirmed the value of learning from normal work, and the power of equipping teams with the tools and mindset to do just that.

As Domtar continues to evolve, the focus remains clear: understanding the reality of work, listening to those who do it, and using what is learned to drive meaningful, sustainable improvement.

DeWayne Bone is the Senior Director of Corporate Health and Safety at Domtar, a leading manufacturer of fiber-based products. In this role, he oversees safety initiatives across Domtar's operations, ensuring the well-being of employees and the continuous improvement of safety practices. Beyond his responsibilities at Domtar, Bone serves as the Ex-Officio Chair of the Board of Directors for the Pulp and Paper Safety Association (PPSA), reflecting his commitment to advancing safety standards within the industry.

About Domtar

Domtar is a leading provider of fiber-based products, including communication, specialty, and packaging papers. With a strong presence in North America, Domtar operates several pulp and paper mills and is known for its commitment to sustainability, operational excellence, and workplace safety.

NATS: AVIATION (UK)

From 7% to 75% Improvement in Risk Reduction: How the UK National Air Traffic Services (NATS) organization implemented proactive safety

Background:
A High-Performing System That Needed More

By the early 2000s, NATS (UK's leading air traffic control provider) was performing very well. It had never experienced a fatal accident, and its safety management system (SMS) was robust. The culture was strong, and safety performance was steadily improving by 6–7% annually.

Yet, when Paul Baron became CEO, he challenged the executive team with a transformative question: *"What would a breakthrough look like?"* for Safety, Service, Value and People. For safety, the answer was clear, eliminate the most serious Safety Significant Events (SSEs). SSEs were a leading indicator of the potential for a mid-air or ground collision. Even in a high-performing system, NATS recognized that being "very good" wasn't enough. The goal became ambitious: zero serious incidents, regardless of who caused them.

Reframing the Problem: From Reactive to Proactive

Rather than continuing to improve mainly based on learning from incidents, NATS made a bold commitment to proactive safety. The mission was to eliminate mid-air collisions *no matter the source, i.e., air traffic controller, airline pilot, private pilot, contractors, airport crew, etc.*

This shift led to a fundamental mindset change: instead of asking "what went wrong?" we asked "how can we support people in doing things right , every time?"

The 'Everybody Knows' Campaign: Listening to the Frontline

The initiative began with an organization-wide campaign called "Everybody Knows." Everyone, from frontline engineers to HR and board members, was asked, "What do you think are the biggest risks of a mid-air or ground collision?" We cross-referenced perceived risks with statistical data and identified our key risk areas.

Each of these risks was then broken down into safety-critical tasks. The focus turned to *task performance*: What were the exact actions that had to go right each time?

A Case in Point: Airspace Infringements

One of the top risks identified was airspace infringement, where unauthorized aircraft (often private or less equipped) entered controlled airspace near busy airports like Heathrow.

We broke the issue down into four critical tasks:

- **Plan** the flight correctly,
- **Navigate** accurately during the flight,
- **Detect** deviations quickly,
- **Avoid** collisions effectively.

Then came a collaborative brainstorm. Ideas flowed, better planning tools for pilots, real-time navigation aids, new alert systems, and even "defensive controlling" that looked beyond the edges of controlled airspace. Controllers were trained to take quicker evasive action when needed, reinforcing that safety always trumped efficiency.

A holistic total system perspective helped us work across organizational boundaries to improve the system to better support people. For example, airspace maps held by NATS were used to enhance pilot planning and navigation tools, and controllers were given better information about pilots flying outside controlled airspace so they could detect issues earlier.

Another Example: Level Busts

Another major issue was pilots flying at incorrect altitudes. Over just 10 days, controllers documented 1,100 cases where pilots were not at the correct level. 99.99% were successfully corrected by the controllers but the data showed that for every 1,000 issues, 1 was not captured by both the pilot and the air traffic controller.

That single miss highlighted the danger. NATS used this data to justify investment in a downlink system, where pilots' flight control settings would be transmitted to controllers, and automatically flagged if they didn't match instructions.

Building a Strategic Safety Plan

These examples weren't isolated efforts. NATS built a full **Strategic Safety Plan** that listed top risks, critical tasks, and intervention strategies. Each safety project had measurable goals, such as "improve detection by 70%." Changes in equipment, training, or procedures were tested in simulators and iterated until they showed measurable improvement.

This structured, data-informed, proactive approach produced stunning results. Instead of improving at 6–7% per year, NATS achieved a **75% risk reduction in just three years**.

75% reduction in 3 years during period of 30% cost savings

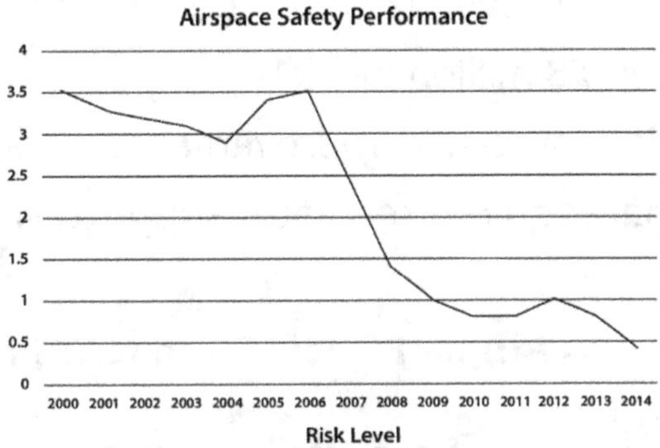

Challenges and courage to focus on proactive

The transformation wasn't without its challenges. Setting a target of *zero* drew skepticism. Some feared it was unrealistic and demotivating, whilst others argued it was immoral to set a target where any number of fatalities would be acceptable. Overall, it was found that aiming for zero forced creative, systemic thinking, asking what would *really* be needed to eliminate risk, not just reduce it.

Another challenge: justifying investment when no incident had occurred. Pulling controllers out of real-time operations, especially in a 24/7 high-pressure environment, wasn't easy. However, leadership alignment made the difference. Goals came from the operations and safety leaders, not imposed from the top. When improvement projects needed key people, the Operations Director personally intervened to make it happen.

Moreover, NATS made a bold financial decision: halving their £8 million incident investigation budget and reinvesting it into systemic improvements identified in previous events. That required regulator support, and courage.

A Ripple Effect: From NATS to Global Aviation

The ethos of proactive learning continued to evolve. When I later joined the UK Civil Aviation Authority (CAA) as Group Safety Director, I helped develop performance-based regulation and introduced the idea that regulatory assessments should focus on actual risk reduction and system performance in addition to compliance with prescriptive rules.

This concept eventually went global. ICAO (International Civil Aviation Organization) adopted the idea into **Annex 19**, requiring each country to create a **State Safety Program** and Plan, a collaborative improvement strategy grounded in proactive risk management.

The influence continued into the offshore oil and gas industry through the work of HeliOffshore. My team developed the **Safety Performance Model** and "**Accident Prevention Goals**," which inspired companies to innovate around critical tasks in order to prevent the next accident.

This model was used industry-wide to:

- agree on areas of focus,
- identify key actions to prevent risk, and
- promote data sharing to make the business case and measure whether actions were having the desired effect on frontline task performance, leading to risk reduction.

In 2025, the model integrated Learning from Normal Work techniques.

Gretchen Haskins is a board member for HeliOffshore Ltd., an organization dedicated to global aviation safety. She is an aviation industry leader in safety performance improvement and an internationally-recognized expert in human factors. Previously, she was CEO of HeliOffshore and before that served on the board of the UK Civil Aviation Authority as Group Director of Safety, overseeing aviation safety in the UK. Prior to that, Gretchen was Group Director of Safety at the UK air traffic company NATS, where she championed activities to support frontline operational safety.

Gretchen has worked in nuclear certification and safety of intercontinental ballistic missiles, joint airworthiness trials for military aircraft, defense system design, and as an expert advisor to NATO on human performance and safety critical systems. She has a strong aviation background and has flown jet and piston aircraft in the U.S. Air Force. She is a graduate of the US Air Force Academy and the University of Southern California in Human Factors Engineering.

About NATS

NATS (National Air Traffic Services) is the main provider of air traffic control services in the United Kingdom, responsible for managing aircraft movements in UK airspace and at several major airports to ensure safe and efficient travel.

PHARMACEUTICALS (International)

Embedding HOP in Contrasts: Lessons from Two Journeys

A Dual Lens on Transformation

Implementing Human and Organizational Performance (HOP) across two vastly different organizational landscapes has been one of the most revealing experiences of my career. One was a global pharmaceutical company, established, structured, and resource-rich. The other, a small biotechnology firm, lean, agile, and still in the process of defining its systems. Despite their differences, both environments highlighted critical truths about what it takes to embed HOP meaningfully.

Why We Chose HOP

In both cases, our starting point was the same: a recognition that traditional approaches for safety and quality often failed to address deeper systemic issues. Mistakes were framed as individual failures rather than opportunities to understand context, constraints, and the reality of work. We needed a way to support our teams more effectively, not through blame, but through understanding. That's where HOP offered a better path forward.

Two Environments, Two Strategies

In the Large, Mature Organization: Integrating with Intent

At the pharmaceutical company, implementing HOP required working within an already intricate web of processes and approvals. Initially, it was a challenge to gain attention. New thinking had to earn credibility in a culture accustomed to "if it isn't broken, don't fix it."

However, once leadership began to see how HOP aligned with our core values, especially our commitment to people, the shift gained traction. We didn't overhaul the system overnight. Instead, we integrated HOP principles into what already existed: enhancing investigations, reframing performance conversations, and using our scale to build momentum.

What made the difference?

- Credibility through Consistency: Integrating HOP into established routines helped reduce resistance.
- Amplification through Reach: With broad organizational influence, small changes cascaded quickly.
- Empowerment through Capability: Deep technical knowledge and mature systems became vehicles for embedding new mindsets.

In the Small, Immature Organization: Building from the Ground Up

The small biotechnology company presented a different challenge: there wasn't much to integrate into. Processes were nascent, capabilities were still forming, and resources were stretched thin. But therein lay the opportunity.

We approached HOP not as a bolt-on, but as a foundation. From defining what "good" looked like to designing systems from first principles, everything was crafted with HOP thinking embedded from the start. Without layers of bureaucracy, we could move fast, if we could make the value clear.

What worked in this environment?

- Simplicity and Relevance: With few legacy systems, our solutions had to be practical and directly beneficial.
- Speed through Sponsorship: When we found the right leaders, change accelerated quickly.
- Curiosity over Conformity: With fewer ingrained habits, teams were open to new ways, if they made sense.

Lessons We Carry Forward

Across both organizations, a few insights held true:

- **Tailor to the Context:** Integration works where maturity exists; implementation works where it doesn't. One size doesn't fit all.
- **Start with People, Not Tools:** The tools matter less than the mindset. When people understand the "why," adoption becomes natural.
- **Build from Stories:** Internal success stories were key in both journeys. Hearing directly from peers about what changed made HOP real.

Embedding HOP is not an endpoint, it's a continuous evolution. Both experiences showed that when we move from blaming to learning, we create space for people to thrive. Whether we're working with legacy systems or building from scratch, the principles remain the same: people are the solution, not the problem.

By staying anchored in that belief, we continue to shift how work is done, how learning happens, and how people are supported, one conversation, one system, one insight at a time.

John Wilkes is currently the Senior Director for Business Excellence working to build and apply Human and Organizational Performance concepts at Novavax. He has more than 30 years of experience within the pharmaceutical and medical device industries at companies such as BioReliance, MesoScale Diagnostics, and AstraZeneca. John's experience was built through working in roles including manufacturing operations, operational excellence, quality systems, quality assurance, and quality control.

UK P&I: SHIPPING (International)

Proactive Learning in the Shipping Industry

The shipping industry, with its intricate operations and global reach, is a sector where safety and efficiency are paramount. As someone deeply invested in this field, I believe that proactive learning is not just beneficial but essential for the industry's sustainable growth and safety.

Proactive learning, particularly through initiatives like "Learning From Normal Work," plays a crucial role in enhancing safety and operational efficiency. This approach involves examining everyday activities on board to identify potential improvements before incidents occur. By focusing on normal operations, we can uncover subtle inefficiencies and risks that might otherwise go unnoticed until they result in significant problems.

On board ships, where the stakes are high, this proactive stance helps in preventing incidents and improving overall performance. It shifts the focus from reactive measures, which often come too late, to a more forward-thinking strategy that aims to mitigate risks before they manifest.

As a leading P&I liability insurer, meeting these challenges is essential. The increasing complexity and size of maritime operations mean that the potential for large-scale incidents is ever-present.

These challenges underscore the need for innovative approaches to risk management. Traditional methods are no longer sufficient in an industry that is constantly evolving. This is where proactive learning becomes invaluable.

The concept of “Learning from Normal Work” resonates deeply with me because it aligns with the proactive mindset necessary for effective risk management. This approach, which has roots in the oil & gas industry, emphasizes understanding the reality of day-to-day operations on board a ship and identifying discrepancies between work as-imagined and work as-done. By doing so, it helps in pinpointing areas where safety and efficiency can be enhanced.

The collaboration between the UK Club and Dr Marcin Nazaruk highlights the commitment to this forward-thinking approach. It reflects a recognition that understanding normal operations can lead to significant improvements in safety culture and operational practices.

Adopting proactive learning strategies like “Learning from Normal Work” brings immense value to the shipping industry. It fosters a culture of continuous improvement and vigilance amongst the crew, where potential issues are addressed before they escalate into major incidents. This not only enhances safety but also boosts operational efficiency and reliability.

By systematically monitoring and analyzing normal operations, Members of the Club can develop more robust risk management frameworks. This proactive stance can improve crew moral, lower operational costs, and build a stronger reputation for safety and reliability in the industry.

In conclusion, proactive learning is a cornerstone of modern risk management in the shipping industry. "Learning From Normal Work" exemplifies how this approach can be effectively implemented, offering a pathway to safer and more efficient maritime operations. As we continue to navigate the complexities of this industry, embracing proactive learning will be key to our success and sustainability.

Stuart Edmondson is the Director of Safety & Risk Management at the UK P&I Club. Before joining the UK Club, he worked as a casualty investigator for a leading shipping law firm, and has extensive experience at sea on various ship types, including crude oil tankers, freight ferries, passenger ships, and offshore drilling units.

About UK P&I

The UK P&I Club provides insurance coverage for shipowners, operators, and charterers against third-party liabilities. This includes risks like damage to cargo, environmental pollution, crew injuries, and other liabilities arising from maritime operations, covering the costs of claims made against the insured parties due to incidents occurring during their shipping activities.

ENERGY INSITUTE (International)

Proactive Learning at the Energy Institute

In 2010, the EI embarked on a transformative journey in incident learning, funded by the Shell-developed Hearts and Minds fund. This decade-long research, led by Professor Alison Littlejohn, Dr. Dane Lukic, and Dr. Victoria Murphy, revealed that organizational learning required more than just mass communication. Incident alerts were often unread and deemed irrelevant, highlighting the need for reflective learning. Reflective learning involves workers and supervisors discussing the implications of incidents and considering how they affect their own work practices. This approach requires time, space, and organizational support, helping teams understand if their work carries similar risks to past incidents and encouraging open discussions. Supervisors use open questions to facilitate these conversations, ensuring lessons are relevant and actionable.

However, learning from incidents doesn't have to be reactive. In 2019, the EI created a website called Toolbox, which hosts lessons learned from incidents across various organizations. This resource allows teams to quickly find an example incident that occurred in another company and use this content proactively before starting a job to understand how they can improve their own work.

Proactive learning extends beyond incident analysis. In 2010, I was involved in developing the publication EI 3201 'Guidance on Safety Critical Task Analysis' (SCTA), which has become one of the EI's most-used documents. Qualitative Human Reliability Analysis (QHRA), sometimes called a human HAZOP, is a formal and robust way of managing human factors risk proactively. The SCTA process involves identifying the most serious hazards and risk scenarios, the tasks that can release those hazards, and the key human actions that, if done incorrectly, will release those hazards. We then carry out a QHRA to identify performance-influencing factors that increase the likelihood of errors and suggest changes to reduce the risk of error to as low as reasonably practicable. Despite its benefits, SCTA is resource -intensive, prompting companies to prioritize tasks for this method. To streamline the process, the Task Improvement Process was developed, offering a quicker, less rigorous alternative that still identifies performance influencing factors and improvements.

In 2025, the EI published EI 3579 'Guidance on Human Factors in Task-Based Risk Assessment,' integrating QHRA into existing risk management tools like job safety analysis. This approach builds the capabilities of work planners and supervisors, enabling them to identify error traps as they identify hazards.

Proactive learning isn't limited to formal risk assessments. Toolbox talks and job planning meetings are excellent opportunities for this. In 2025, the EI introduced the Energy and Error Wheel, a simple diagram that helps teams bring QHRA even closer to home, to discuss hazards and error traps together before and after a job, fostering continuous improvement.

Moreover, in 2020, the EI published 'Making Compliance Easier,' a workshop-based activity aimed at improving procedures and the work environment. This tool helps companies identify and fix error traps and other issues without needing an incident first.

A key takeaway is that proactive learning can be formal, captured in changes to tasks, procedures, systems, policies, equipment, and environment. But it can also be informal, captured in the hearts and minds of individuals, and for this reflective learning is vital. Both are important and neither should be neglected.

Stuart King has worked for the EI since 2009, where he manages the EI's activities related to human and organizational factors. This work has been developed by various human factors subject matter experts, and primarily overseen by the EI's Human and Organizational Factors Committee. Together, this group of stakeholders has, since 2002, helped the EI to publish nearly 80 resources on human factors, as well as other resources, many of which are made freely available to the public, and many of which aim to help organizations learn proactively.

About The Energy Institute (EI)

The Energy Institute (EI) is a professional organization for individuals working in energy-related fields. The EI was formed in 2003 by the merger of the Institute of Petroleum (dating back to 1913) and the Institute of Energy (dating back to 1925). It has an international membership of about 20,000 people and 200 companies. It has a technical and innovation work program, where it develops and publishes industry good practice on a range of topics related to health, safety and environmental management.

OIL AND GAS (International)

The shift towards Learning from Normal Work

Every business knows that to remain relevant and successful in today's dynamic world of technological innovation and global interconnections, organizations must encourage continuous learning to stay ahead of potential challenges.

Traditionally, especially in health and safety, we assumed everything was fine when no negative incidents occurred and waited for adverse outcomes to drive improvements. Over the past decades, learning from incidents and advancing human and industrial safety have successfully reduced negative occurrences, seemingly diminishing opportunities for further learning.

However, not all errors lead to failure, and not all failures result in negative outcomes. In fact, most organizations gradually improve until they reach a performance plateau, often experiencing a phase of learning drought.

This is where Learning From Normal Work can make a significant difference. It involves proactively learning from day-to-day work that does not result in negative outcomes or failures. Normal work is neither good nor bad, safe nor unsafe; it represents the 'normal' performance variation where people adapt and adjust daily to meet demands for time, quality, and cost despite constraints, limitations, and obstacles.

Baker Hughes

At Baker Hughes, Learning From Normal Work has helped us shift the focus from reacting to past failures to anticipating future risks based on real operational knowledge. It has helped surface insights that traditional risk assessments often miss, making it a powerful method for identifying hidden vulnerabilities in normal work and prompting meaningful system improvements.

International Association of Oil and Gas Producers (IOGP)

By leveraging the International Association of Oil and Gas Producers (IOGP) Report 642, 'Learning From Normal Work', the first major global publication to give structured recognition to proactive learning from normal operations, we elevated the credibility of proactive learning among executives and regulators. Alongside us, many other companies found it to be a practical starting point, addressing not just 'why' we should learn from normal work, but 'how'.

This is where, at Baker Hughes, we leveraged the Finding The Next Incident approach beside the well-known tools of Learning Teams and Walk-Through Talk-Through (WTTT). This approach involves encouraging employees to envision potential serious incidents and describe their details within their operational areas. By tapping into their operational insights, experience, and imagination, we effectively harnessed their ability to foresee possible interactions of events and causal factors throughout the chain of activities, enabling them to identify and address hidden risks.

Finding The Next Incident (FTNI) has shifted the conversation toward Serious Injuries or Fatalities (SIF), particularly those resulting from the failure of complex systems, where the level of risk becomes apparent only when a specific combination of events is triggered. But it has also enhanced leadership resilience and created a psychologically safe environment for employees to contribute to the safety journey.

Energy Workforce & Technology Council (EWTC)

All these tools have been recently included in "The Human and Organizational Performance Playbook" (2025), co-authored with Abdelsalam Yasseen for the Energy Workforce & Technology Council (EWTC) – a guide that serves as a foundational resource for companies ready to take their first steps in implementing Human and Organizational Performance (HOP) and LFNW. In this playbook, we clarified the difference between Work as Imagined vs. Work as Done and presented Learning Teams, WTTT, and FTNI – with detailed guidance on structure, time, participants, and outcomes.

Center for Offshore Safety (COS)

This is also reflected in the activities of the "Learning From Normal Work" workgroup within the Center for Offshore Safety (COS) of the American Petroleum Institute, where I've had the privilege to help shape a structured conversation around proactive learning in the offshore sector. Our group focuses on integrating Learning From Normal Work into safety practices. This involves moving away from waiting for an incident to happen before we learn, and instead designing ways to learn proactively from how work is actually done by frontline personnel, in real conditions, facing real constraints.

International Association of Drilling Contractors (IADC)

Finally, the recognition that Learning From Normal Work is being embraced by the industry can be clearly seen and heard during the insightful conferences of the International Association of Drilling Contractors (IADC). IADC plays an important role in advancing proactive learning across the drilling industry by fostering open dialogue and knowledge exchange through its global conferences, the work of its technical committees, its publications, the accreditation program, and not last, advocacy. The focus has increasingly shifted toward learning from normal work, integrating Human Performance principles, and building a culture that values insights from frontline operations before incidents occur.

The oil and gas industry is investing significantly in this subject, recognizing the immense value of proactive learning. By embracing Learning from Normal Work, we can move beyond compliance and towards continuous learning and operational resilience.

Learning From Normal Work is the frontier where proactive learning becomes possible; this book is a call to action to move from systems that respond to failure to systems that learn continuously from what goes right.

Lamberto Nonno is the Director of HSE Governance and Customer Excellence at Baker Hughes. He chairs the Human Performance Committee at the IOGP, and the "Learning from Normal Work" workgroup at the Center for Offshore Safety; Lamberto has also co-authored the Energy Workforce & Technology Council's HOP playbook and he's playing a key role in shaping the industry's shift towards learning from normal work and systematic safety improvement. With more than two decades of experience in the industry and a lifelong passion in psychology and sociology, Lamberto's journey is one of relentless pursuit in advancing our view of human potential.

The views expressed are those of the author and do not necessarily reflect the views or positions of the aforementioned associations or their members.

Learning From Normal Work glossary

(Terms in order of appearance in the story)

Five Whys

The *Five Whys* technique is a problem-solving method that encourages investigators to look beyond the immediate cause of a problem or accident, and to explore the root causes of an adverse event. The technique is simple: ask 'why?' five times, with each 'why' challenging the previous answer, and each answer leading closer to finding the root cause.

In John's story, *Five Whys* is used to trace the cause of the accident in chapter one. As the story shows, this technique does not always produce a satisfactory result. While it can provide a helpful starting point, investigators must make sure that they have asked enough 'why' questions to uncover the fundamental causes of the problem.

Normal Work

Normal work is the reality of how a job gets done day to day. It is not just the written procedure or the official plan, but the way people actually carry out the task in real conditions. Workers often face things that make the job harder: missing tools, unclear instructions, time pressure, equipment that doesn't fit, or unexpected changes. To get the job done, they adjust, improvise, and come up with practical solutions on the spot.

Because of this, normal work is full of small challenges and adaptations. These adjustments are not signs of people ignoring the rules; they are usually what allows the job to move forward. By paying attention to normal work, leaders can see the gap between "how the job is supposed to be done" and "how it really happens." This gives valuable insights into what creates risk, what slows people down, and what can be improved to make work safer and easier.

Learning From Normal Work

Learning from Normal Work is a scalable framework that helps large organizations move beyond reacting to accidents and instead learn proactively from the everyday realities of work. Instead of waiting for incidents to reveal vulnerabilities, the approach focuses on understanding how tasks are actually carried out, what makes them difficult, and where systems or processes create hidden risks.

The framework recognizes the role of different stakeholders and provides relevant tools for each level, for example, strategic approaches for senior leaders; tactical, hands-on tools for supervisors; or in-depth techniques for safety and quality professionals, making it practical to implement across the entire organization.

HOP (Human and Organizational Performance)

Human and Organizational Performance is an umbrella term for many of the approaches John uses in the story to improve Health and Safety management. It includes understanding that humans make mistakes, and that blame does not eliminate the root cause of an accident. The HOP philosophy recognizes that elements of the work environment can lead to error traps, and that a thorough understanding of the workplace and the work as-done processes involved is essential to improving the day-to-day safety of employees.

John and Karen discuss HOP approach in chapter two, and the concept is used in various contexts throughout the story.

Error Traps

An error trap is anything that could make a human operator more likely to make a mistake or not follow the rules. This could include unclear procedures, confusing design, or unavailability of the correct tools.

Learning Teams

A Learning Team is a group conversation with the people who do the job every day. The purpose is to hear their stories about how the work really happens, what goes smoothly, what gets in the way, and what they have to adjust to make things work.

It works best when everyone feels comfortable sharing openly, and when the focus stays on making the work easier, safer, and more effective.

In practice, it usually looks like this:

1. A facilitator brings people together after an event or just to learn from normal work.

2. The group talks through the process, sharing where things go well and where they struggle.

3. The insights are written down and turned into actions that help reduce problems and build on what already works.

IOGP (International Association of Oil and Gas Producers)

The International Association of Oil and Gas Producers published an industry guide to Learning From Normal Work, mentioned by John at his team meeting in chapter five.

TEDS (Tell, Explain, Describe, and Show)

Tell, Explain, Describe, and Show (TEDS) is a technique for asking open questions, which encourages employees to open up and share their experiences and observations. It enables open and honest conversations, without implying fault or blame.

In John's story, TEDS is demonstrated by Sarah in chapter five. She asks employees on the factory floor to *Tell* her about the key challenges they face with a piece of equipment, to *Explain* more about the problems, to *Describe* an incident that was difficult to deal with, and to *Show* her the problem. She follows these questions and answers with more TEDS questions in order to fully understand the employees' concerns.

Walk-Though Talk-Through

A Walk-Though Talk-Through (WTTT) allows a manager or supervisor to gain an understanding of the real-life processes experienced by their employees. In order to fully understand a process, the supervisor invites an operator to walk them through every step of the process, explaining each element of the task as they go. The supervisor has the opportunity to ask about potential problems and errors at each stage, and to discuss potential solutions to these issues.

John uses the *Walk-Though Talk-Through* technique during the pilot project in chapter nine.

HSE

HSE stands for Health, Safety, and the Environment.

John co-ordinates with HSE managers and safety specialists throughout the story.

Work-As-done

How tasks are actually carried out in real conditions, with adaptations and adjustments workers make to complete the task. (See: Normal Work)

Work As-imagined

How tasks are expected to be carried out, typically described in procedures, plans, or management assumptions.

Safety II

Safety II is an approach to Health and Safety management that focuses on processes that go right, rather than on things that go wrong. As well as helping to avoid safety incidents, Safety II helps organizations to learn, anticipate problems, and continue to improve their Health and Safety policies.

Just Culture

In contrast to a 'blame culture', where individuals may be blamed for mistakes made due to systemic issues, a Just Culture promotes openness and learning from mistakes. Rather than hiding risky behaviors, workers are encouraged to report errors, and the organization aims to learn from their reports.

Learning From Normal Work Competency Pathway

The "Learning From Normal Work Competency Pathway" is a practical and structured program designed to help organizations build internal expertise in proactive safety and operational learning. It's aimed at people whose job is to bring modern safety thinking to life inside the company or support proactive learning rollout efforts.

The pathway includes twelve modules that cover key areas such how to proactively identify error traps, how design sets people up for failure, how to predict human error or non-compliance, psychology of decision making, how to integrate with risk assessment, investigations, accountability or KPIs.

It provides practical tools like checklists, and interview guides that can be used immediately in daily work.

The Pathway is accredited by the International Society for Performance Improvement (ISPI), a professional association that helps individuals and organizations improve results by applying evidence-based practices in human performance, learning, and organizational development.

About Psychology Applied

Psychology Applied Limited (PAL) is a training and consultancy company led by Dr Marcin Nazaruk, specializing in enabling practitioners and large organizations to implement Learning From Normal Work at scale.

PAL offers a range of training and integration packages:

- **Mindset Shifter**: 1-day workshop for leaders to learn the basics of performance variability, adaptive behaviors, error traps and practical tools.
- **Planning Workshop**: 1-day planning workshop to co-create an implementation plan.
- **Competency Pathway**: 5-day face-to-face or 6 months (online-blended), online, highly practical, in-depth training for internal champions to upskill safety, quality and operational personnel in a range of proactive learning tools and techniques.

- **Context Finder**: 1-day workshop for front-line supervisors to teach hands-on learning techniques that can be immediately applied in the front line.
- **Trap Spotter**: 0.5-day training for the workforce to help identify error traps.
- **Awareness e-learn**: 2h e-learning to help support functions understand basic concepts and tools.
- **Train-the-trainer**: 2-day workshop to develop internal trainers capable of delivering 'Context Finder' for supervisors.
- **Bespoke Implementation Support**
- **Additional Tools**:
    a. Learning from Normal Work Discussion Cards
    b. Tally books for supervisors and team leaders
    c. AI Coach to learn questioning skills
    d. AI Feedback on the quality of learning

For companies wishing to upgrade their existing programs, PAL offers a Re-Imagine series focused on tweaking and adjusting existing processes to learn more deeply:

1. Re-Imagine Risk Assessment
2. Re-Imagine Behavioral Observations

3. Re-Imagine Accident Investigations

Find out more at **learningfromnormalwork.com**

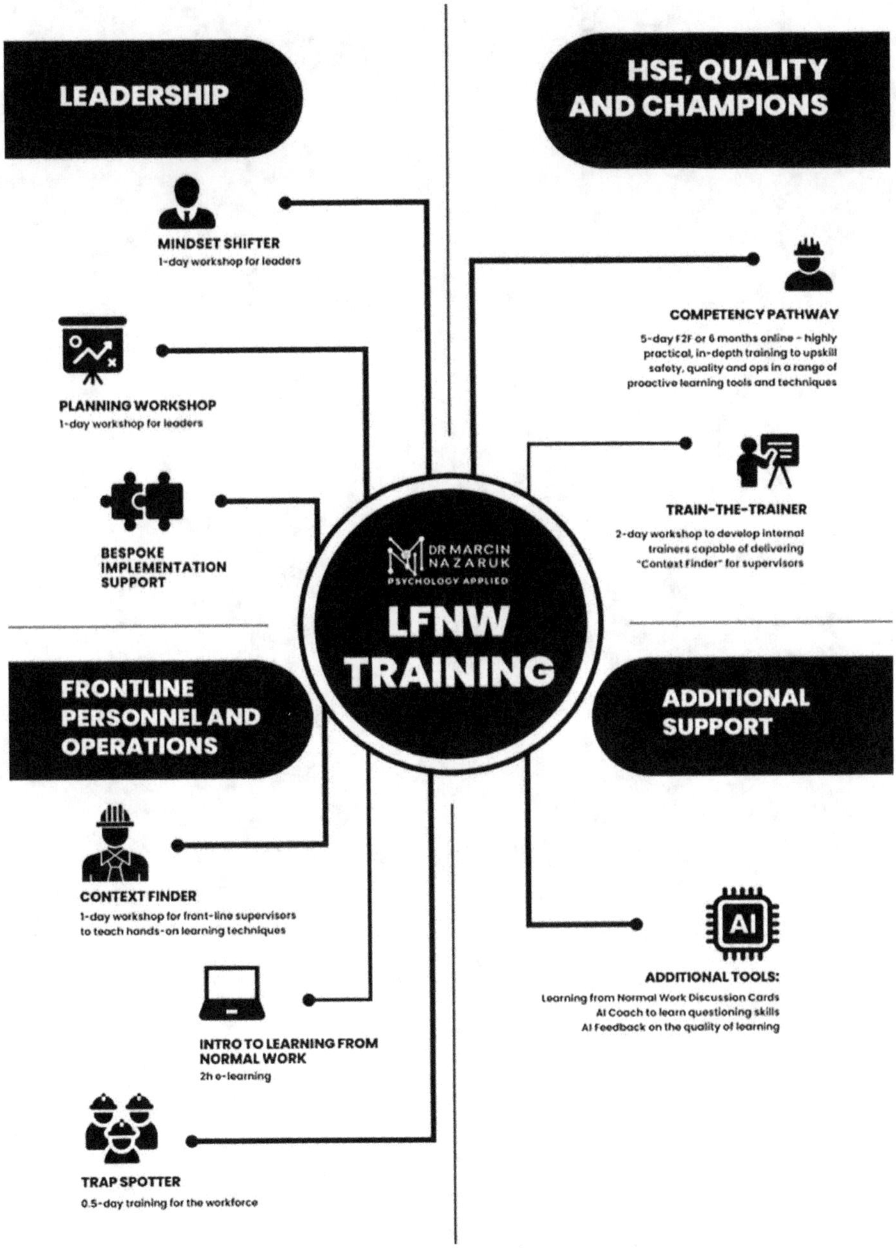

About the author

Dr Marcin Nazaruk is a globally recognized expert in safety transformation, renowned for bridging the gap between academic research and practical application.

With a PhD in the industrial psychology of safety transformation, and over two decades of hands-on experience, he has led global initiatives focused on enterprise-wide implementation of proactive learning at major organizations.

Dr Nazaruk has authored multiple industry guidance documents, including the IOGP Guide 642 on Learning From Normal Work, and has developed practical tools that have been widely adopted across the sector.

His contributions have earned him multiple awards, including the SHP Trailblazer Award and the SPE International HSE Award.

As the founder of Psychology Applied, he continues to help large organizations implement proactive learning strategies to enhance safety and operational performance.

For more information, visit
www.learningfromnormalwork.com

Scientific foundations underpinning learning from normal work

Albrechtsen, E., Solberg, I., & Svensli, E. (2019). The application and benefits of job safety analysis. *Safety Science, 113*, 425–437.

Arezes, P. M., & Sérgio Miguel, A. (2003). The role of safety culture in safety performance measurement. *Measuring Business Excellence, 7*(4), 20–28.

Barnett, R. L. (2020). On the safety hierarchy and hierarchy of controls. *American Journal of Mechanical Engineering, 8*(2), 61–68.

Bentley, S. K., McNamara, S., Meguerdichian, M., Walker, K., Patterson, M., & Bajaj, K. (2021). Debrief it all: a tool for inclusion of Safety-II. *Advances in Simulation, 6*(1), 1–6.

Bieder, C., & Bourrier, M. (2013). *Trapping Safety Into Rules: How Desirable Or Avoidable Is Proceduralization?* Ashgate Publishing, Limited.

Blackman, H. S., Gertman, D. I., & Boring, R. L. (2008). *Human Error Quantification Using Performance Shaping Factors in the SPAR-H Method*. Idaho. Retrieved from https://inldigitallibrary.inl.gov/sites/sti/sti/4074955.pdf

Boskeljon-Horst, L., De Boer, R. J., Sillem, S., & Dekker, S. W. A. (2022). Goal conflicts, classical management and constructivism: How operators get things done. *Safety, 8*(2), 37.

Boyce, T. E., & Geller, E. S. (2001). Applied behavior analysis and occupational safety: The challenge of response maintenance. *Journal of Organizational Behavior Management, 21*(1), 31–60.

Cadieux, J., Roy, M., & Desmarais, L. (2006). A preliminary validation of a new measure of occupational health and safety. *Journal of Safety Research, 37*(4), 413–419.

Conklin, T. (2016). *Pre-Accident Investigations: Better Questions - An Applied Approach to Operational Learning*. CRC Press.

Cook, R. I. (1998). How complex systems fail. *Cognitive Technologies Laboratory, University of Chicago. Chicago IL*, 64–118.

Dekker, S. W. A., Long, R., & Wybo, J.-L. (2016). Zero vision and a Western salvation narrative. *Safety Science, 88*, 219–223.

dos Santos, I. J. A. L., França, J. E. M., Santos, L. F. M., & Haddad, A. N. (2020). Allocation of performance shaping factors in the risk assessment of an offshore installation. *Journal of Loss Prevention in the Process Industries, 64*, 104085.

EuroControl. (2014). *Systems Thinking for Safety: Ten Principles A White Paper. Moving towards Safety-II*. Retrieved from http://skybrary.aero/bookshelf/books/2882.pdf

Flight Safety Foundation. (2021). *Learning From All Operations: Expanding the Field of Vision to Improve Aviation Safety.*

Gerhart, B. A., & Rynes, S. (Sara). (2003). *Compensation : theory, evidence, and strategic implications*. Sage.

Grant, E., Salmon, P. M., Stevens, N. J., Goode, N., & Read, G. J. (2018). Back to the future: What do accident causation models tell us about accident prediction? *Safety Science, 104*, 99–109.

Hale, A., & Borys, D. (2012). Working to rule, or working safely? Part 1: A state of the art review. *Safety Science*, 1–15. https://doi.org/10.1016/j.ssci.2012.05.011

Hendricks, J. W., & Peres, S. C. (2021). Beyond human error: An empirical study of the safety Model 1 and Model 2 approaches for predicting workers' behaviors and outcomes with procedures. *Safety Science, 134*, 105016. https://doi.org/10.1016/J.SSCI.2020.105016

Hollnagel, E. (2002). Understanding accidents-from root causes to performance variability. In *Proceedings of the IEEE*

7th Conference on Human Factors and Power Plants. IEEE. https://doi.org/10.1109/HFPP.2002.1042821

Hollnagel, E. (2017). *Safety-II in Practice: Developing the Resilience Potentials*. CRC Press.

Hollnagel, E., Leonhardt, J., Licu, T., & Shorrock, S. (2013). From Safety-I to Safety-II: a white paper. *Brussels: European Organisation for the Safety of Air Navigation (EUROCONTROL).*

Homann, F., Limbert, C., Bell, N., & Sykes, P. (2022). Safety through engaged workers: the link between Safety-II and work engagement. *Safety Science, 146*, 105521.

Hughes, P., & Ferrett, E. (2011). *Introduction to Health and Safety at Work: The Handbook for the Nebosh National General Certificate*. Butterworth-Heinemann.

IOGP. (2022). *Learning from Normal Work.*

Kirwan, B., & Ainsworth, L. K. (2003). *A Guide To Task Analysis: The Task Analysis Working Group*. Taylor & Francis.

Kirwan, B., Bettignies-Thiebaux, B., Matteo, C., Baumler, R., & Carrera Arce, M. (2020). *Towards a Safety Learning Culture for the Shipping Industry.*

Lindberg, A.-K., Hansson, S. O., & Rollenhagen, C. (2010). Learning from accidents–what more do we need to know? *Safety Science, 48*(6), 714–721.

Liu, J., Zou, Y., Wang, W., Zhang, L., Liu, X., Ding, Q., ... Čepin, M. (2021). Analysis of dependencies among performance shaping factors in human reliability analysis based on a system dynamics approach. *Reliability Engineering & System Safety, 215*, 107890.

Pan, X., & Wu, Z. (2020). Performance shaping factors in the human error probability modification of human reliability analysis. *International Journal of Occupational Safety and Ergonomics, 26*(3), 538–550.

Rozenfeld, O., Sacks, R., Rosenfeld, Y., & Baum, H. (2010). Construction job safety analysis. *Safety Science, 48*(4), 491–498.

Sherratt, F., & Ivory, C. (2019). Managing "a little bit unsafe": Complexity, construction safety and situational self-organising. *Engineering, Construction and Architectural Management*.

SPE. (2020). *ARE YOU APPLYING HUMAN FACTORS / HUMAN PERFORMANCE AS PER THE INDUSTRY GUIDANCE?*

Staddon, J. E. (1979). Operant behavior as adaptation to constraint. *Journal of Experimental Psychology: General, 108*, 48–67. https://doi.org/10.1037/0096-3445.108.1.48

Wickens, C. D., Gordon, S. E., Liu, Y., & Lee, J. (2004). *An introduction to human factors engineering* (Vol. 2). Pearson Prentice Hall Upper Saddle River, NJ.

Reviews

"Learning from Normal Work offers a compelling, practical narrative on transforming workplace safety by shifting from blame to learning. Through the fictional yet relatable journey of John, a safety manager, the book illustrates the challenges of moving beyond traditional safety metrics to proactively understand everyday work. It blends storytelling with evidence-informed practices, making concepts like Human and Organizational Performance (HOP) approachable and relevant. Most importantly, it delivers a practical roadmap for organizations transitioning toward a learning culture. This engaging, instructive guide is ideal for safety leaders seeking culture change, reduced risk, and authentic engagement with frontline workers—a vital read for modern safety transformation."

—Andresa Hernandes

Vice President of Safety

Siemens AG

"A relatable journey through the eyes of John, a Senior Safety Director, as he transforms his company's safety mindset. His real-world experiences, triumphs, and setbacks reflect challenges we all face in EHS. Along the way, supportive colleagues who've walked similar paths show how learning from everyday work leads to meaningful change. A refreshingly easy and inspiring read that reminds us how shared knowledge and collaboration can spark new outcomes."

—Donald Lyons

Senior Director, Global Workplace EHS/OH Excellence

Kenvue

"We often think we're doing everything "by the book"—but what if the book doesn't reflect reality? That's one of the key takeaways from Learning From Normal Work. Too many safety systems are built on assumptions about how work should happen, not how it actually does. This book helps close that gap.

Marcin shows how stopping at "human error" or "complacency" might be common—but it's not useful. Complacency isn't a flaw in the worker; it's a feature of the human brain. We're wired to go on autopilot when tasks become routine. We can't fix that with slogans or retraining—we have to design systems that expect it and protect people when it happens.

And what really sets this book apart is how clearly it shows the importance of listening—not just asking questions, but asking the right ones in the right way, and really hearing the answers. At one point, John learns that asking "Why didn't you follow the procedure?" shuts people down. "But when he shifts to the TEDS method—Tell, Explain, Describe, Show—and creates space for workers to speak openly, the insights start to flow.

A personal moment in the story drives this home. John's wife, Melissa, tells him: "Questions are good, but you need to listen to the answers... Give me space to talk, before you make it all about you again." That line could be said to any leader. Listening without judgment builds trust. And when workers trust that you won't weaponize their honesty, they start offering the kind of insight that actually prevents accidents.

This book is more than a framework—it's a reminder that safety happens in relationships. Trust and collaboration aren't soft skills—they're essential tools for building systems that work."

—Thomas Stewart

Director of Safety & Security

Range Resources

"The LFNW book is loaded with well-thought-out, practical and simple-to-follow content. Any safety professional or business leader can learn a lot about the fundamentals of LFNW, its nomenclature and benefits from the story about John. It gives hope that failing safety cultures can turn around from reactive to proactive once there is the motivation and support systems. I

would recommend this book to all safety practitioners who have tried traditional approaches with little or no success but desire to see real change...for the better!"

—Keron Joseph
Head, Process Safety Management
National Energy Corporation of Trinidad and Tobago Limited

"This book is ideal for leaders and safety professionals seeking more knowledge of what Learning From Normal Work (LFNW) is and which benefits and challenges are expected during implementation. The book introduces new views on workplace safety, risk and human performance. By focusing on collaboration and learning from daily tasks rather than blame, this book provides invaluable insights for creating a safer and more resilient workplace. With practical examples and actionable advice, it offers useful strategies to build safer, more effective teams and improve performance. A valuable resource for anyone committed to safe operations."

—Gøran Furland
VP QHSE
Odfjell Drilling

"Dr Marcin Nazaruk has not just written an important book for leaders and practitioners on taking learning and improvement to the next level, he as a practitioner has detailed what it takes to seamlessly embed Learning From Normal Work into your organization.
The book provides insights and is a practical guide for leaders and practitioners who desire to shift from learning only from failure to learning mainly from normal work before experiencing events. Marcin brings a wealth of experience, deep knowledge, and practical ideas to leaders and practitioners everywhere trying to go beyond reactive learning and seeking compliance.
Learning From Normal Work is simple and powerful, taking you from concept to principle to practice in an entertaining and masterfully informative industrial novel. I was taking notes on almost every page and I will definitely be providing this book to my leadership and practitioner teams.
A must-read for any leader or practitioner interested in understanding the game-changing advantages of Learning From Normal Work. This book is a goldmine of ideas and practical steps.... Read it, highlight it, learn from it, and put it into practice. It will change the way you learn, you engage, and your results."
—Charles Major
Director of Operational Excellence and Human Performance at Luminant
President at the Community of Human and Organizational Learning

www.ingramcontent.com/pod-product-compliance
Lightning Source LLC
LaVergne TN
LVHW011522200926
848484LV00001B/119